Mechanical Energy Conversion

Series Editor
Alain Dollet

Mechanical Energy Conversion

Exercises for Scaling Renewable Energy Systems

Mathieu Mory

iSTE

WILEY

First published 2024 in Great Britain and the United States by ISTE Ltd and John Wiley & Sons, Inc.

ISTE Ltd
27-37 St George's Road
London SW19 4EU
UK

www.iste.co.uk

John Wiley & Sons, Inc.
111 River Street
Hoboken, NJ 07030
USA

www.wiley.com

Library of Congress Control Number: 2023952598

British Library Cataloguing-in-Publication Data
A CIP record for this book is available from the British Library
ISBN 978-1-78630-923-5

Contents

Foreword

The subtitle of this work refers to "renewable energies" and it is for this reason that I was asked by Mathieu Mory to write this foreword. My professional life has mostly been spent in the Laboratory of geophysical and industrial flows (*Laboratoire des écoulements géophysiques et industriels* [LEGI]), which is a Joint Research Unit (UMR 5519) of the French National Center for Scientific Research (*Centre national de la recherche scientifique* [CNRS]), the Institut national polytechnique de Grenoble (Grenoble INP) and the University of Grenoble-Alpes (UGA). It was in this laboratory that I launched a research program in 2000, bringing together four regional laboratories over a period of around 10 years, the objective of which was to design and ultimately produce a demonstrator of an innovative marine tidal turbine. This program was at the origin of the company HydroQuest, whose tidal turbine is mentioned in this book. Since then, I have never ceased to be involved in renewable energies, first through several studies on new converters, and at the same time by participating in various national organizations, such as group No. 5 on marine, hydraulic and wind turbine energies of the *Alliance nationale de coordination de la recherche pour l'énergie* (ANCRE) and like the Scientific Evaluation Committee CES 05 – Sustainable, Clean, Safe and Efficient Energy of the French National Agency for Research (ANR).

This book is original in several respects.

First of all, by its angle of attack on renewable energies that are limited to hydraulic and wind energy, that is to say from water and wind, respectively; the three other commonly accepted forms thus being not considered, namely, solar, biomass and geothermal energies, coming from the sun, from the biomass or from the earth. The original limitation appears in fact in the title itself, which implicitly specifies that only mechanical conversion will be considered in the conversion chain of the various sectors covered; thus, the part of the chain taken into account extends from the primary resource available in nature and stops upstream of the electrical generators. On the other hand, this part is very carefully treated, step by step, component by component, some of the latter possibly not being mechanical: thermal in the case of marine thermal energy or chemical in the case of converters of osmotic energy. It should be noted that the division into chapters covers the wind power sector (Chapter 3) and the six hydropower sectors (Chapters 4–8). The division is classic, except that Chapter 4 devoted to tidal energy deals with both the tidal dam sector and the marine tidal turbine sector.

Another original aspect is the introduction of an inventory, carried out with moderation, for each sector considered, appearing either at the beginning of the chapter concerned, or at the end. This inventory is not without interest for students who, at the end of their studies, wish to either join a company or undertake a thesis to develop one of the renewable energy sectors considered here. They may feel encouraged by the media which do not always have the perspective necessary to judge the maturity of the various sectors discussed here, nor the importance of the obstacles to be overcome in order to obtain acceptable costs. The resulting enthusiasm is certainly eminently positive for the decarbonization of our planet, but it must be channeled to truly meet the expectations of these students, so that they are not disappointed. For each sector, it is desirable that they can a priori place in a general context the objectives which will be proposed to them, know the part of the strictly technical or economic requirements to be raised that the innovation makes it possible in priority to overcome, and the part from scientific requirements of a more fundamental nature to optimize sectors that are beginning to be well known and exploited.

The last original aspect and certainly the most important: the approach used to transfer the information transmitted over the chapters based on a pedagogy that can be described as participatory. The succession of exercises provided in each sector in this book, exercises which break down into a succession of questions, followed by the succession of corresponding answers, invites students to think for themselves to discover independently. These exercises relate to situations drawn from well-described applications that become significant for them; they are in a way on a bridge between statements of very theoretical and abstract problems and statements of problems containing too technical details. Said exercises are finally relevant without being too complex: they do not require elaborate numerical approaches. We can imagine that they are themselves the fruit of experiments, trial and error, resulting from lectures, and exercise sessions offered by the author to his students.

To conclude, I would like to indicate the reasons for which I agreed to write this foreword for this book. Mathieu Mory is a friend that I have known since 1982 and I know his many qualities. However, I would be careful not to make a panegyric, but I simply want to communicate the word that comes to mind when I think of him. This word is rigor, moral as well as intellectual rigor. This rigor explains his professional trajectory since leaving the Ecole Polytechnique. He was then offered a post of high responsibility in the R&D department of a large company, which he refused because of a knowledge which he then considered insufficient of the technical-scientific activities of this department. This is how he chose and pursued a university career in which teaching was never disdained in favor of a research activity that remained of a high level. This book is an illustration of that.

Jean-Luc ACHARD
Emeritus Research Director at the CNRS

Preface

Despite all the efforts, it is certain that errors remain in this work, since the last proofreading still reveals some…

When we were working on the exercises in class, the numerical applications were carried out with an Excel spreadsheet. I invite the reader to do the same. This makes it possible to see the effect of the parameters on the result and errors in numerical application are more easily identified.

The results that appear in the exercise tables are copies of the spreadsheet results. If the reader finds different numerical results, they are not necessarily the one who is making an error.

I invite the readers, if they discover an error, if they have a comment or a question, to let me know. Reach out to me at the e-mail address[1] created in connection with this book. I will try to answer as best I can.

Please accept my apologies and thanks in advance.

February 2024

1 mmenergie72@gmail.com.

Acknowledgments

I would like to thank my colleagues Didier Graebling, Daniel Broseta and Jean-Luc Achard for their advice on this manuscript.

To students of the international master's degree in Simulation and Optimization of Energy Systems (SIMOS):

– Alireza Abbasi (Iran);

– Shayan Baghi (Iran);

– Sarad Basnet (Nepal);

– Yacine Bounekta (Algeria);

– Angelica Chacon (Venezuela);

– Mohamed Chouaikhia (Tunisia);

– Arpan Dutta (India);

– Godswill Ighedosa (Nigeria);

– Gholamhossein Kahid Baseri (Iran);

– Mukesh Kumar (Pakistan);

– Shubham Manchanda (India);

– Armin Mokhtari Shargi (Iran);

– Samandar Nabiev (Uzbekistan);

– Christopher Offor (Nigeria);

– Asaph Palencia (Philippines);

– Mohammad Sadr (Afghanistan);

– Karan Shahi (India);

– Ivan Vasilev (Russia).

Without you, without your attention, without your questions, without your presence, this course would not have been developed and this book would not have been written.

In dedicating this work to you, I want to express my friendship in return for that which you have shown me.

Introduction

I.1. Presentation of the book

This book approaches mechanical conversion by proposing a series of exercises and by presenting their solutions. These are generally scaling exercises. They implement methods which determine the characteristics and dimensions of installations and machines which, for the different studied processes, provide the desired power or quantity of energy.

The fields of application affect a variety of marine energies: tidal and hydrokinetic energy, wave energy, marine thermal energy and osmotic energy. They also include wind energy and hydroelectricity. The different areas of application are discussed in separate chapters. For each type of energy, the energy resource is described, a process for recovering it is presented and its dimensioning is carried out for a fixed production objective.

Proposing an exercise is a simplification when compared to the real situation with which the engineer is confronted. We are not required to comply with the constraints of a site or a real situation. Nevertheless, the conditions of the proposed exercises were not chosen by chance. For many of them, we had in view a specific installation and, in return, it was sometimes possible to compare the scaling carried out with known data on these installations. Reference facilities are named when they were used to develop an exercise. The information available is often incomplete. Considering dimensioning as an exercise allows you to keep a distance from a situation that is only partially known. In doing so, the exercise makes it possible to focus on the principles of mechanical conversion which constitute the subject of this book.

The scientific discipline to which these exercises belong is mainly fluid mechanics. This is why Chapter 1 offers reminders in fluid mechanics. It should be seen as a memorandum rather than a lesson. The reader can refer to fluid mechanics books to read the missing demonstrations. This chapter basically provides the elements necessary for the resolution of the exercises. Chapter 2 presents course reminders on hydraulic turbine engines. Two exercises are included in it. Other exercises directly related to the methods exposed in Chapter 2 are found in the following chapters, devoted to wind energy, hydroelectricity or tidal energy.

The fields of application of the exercises fall under renewable energies. This is why the term *renewable energies* appears in the subtitle of this book. For each sector, a short inventory is drawn up in the chapter devoted to it. The objective is for the reader to know the degree of maturity and to have a notion of the powers and energies delivered by the installations currently built. Some of these sectors are rapidly changing. The figures given are those known at the time of publication of this work. They can quickly become obsolete. This is the reason why this inventory remains succinct. This is not the objective of this book, which aims to contribute to the acquisition of dimensioning methods. The reader will find an up-to-date inventory of a sector in the reports published in France by the Renewable Energies Syndicate[1] or by ADEME.

The operation of the electricity production facilities studied in this work, which use renewable energies, do not produce CO_2. Nevertheless, the manufacture of the equipment necessary for the operation of the factories involves processes which in themselves emit CO_2. An analysis of the lifecycle of the installations is necessary to establish the true carbon footprint of these units. This aspect is not covered in this book.

I.2. Power, energy and load factor

Energy data are quantified by energy and power. Power is energy per unit time. In steady state, if an energy system delivers power $\wp$ at a constant rate, the energy transferred during a duration T is calculated by the formula $E = \wp T$.

In the international unit system, power is measured in Watts. The Watt is converted into international units (kg, m, s) according to:

$$1\,\mathrm{W} = 1\,\mathrm{kg\,m^2 s^{-3}}$$

1 "Panorama de l'électricité renouvelable au 30 juin 2021", publication of the French Renewable Energies Syndicate. These publications are annual.

Energy is measured in Joules (J), with the equivalence in units (kg, m, s):

$$1\,\mathrm{J} = 1\,\mathrm{kg\,m^2 s^{-2}}$$

Energy data vary over wide ranges. This is why kiloW ($1\,\mathrm{kW} = 10^3\,\mathrm{W}$), MegaW ($1\,\mathrm{MW} = 10^6\,\mathrm{W}$), GigaW ($1\,\mathrm{GW} = 10^9\,\mathrm{W}$) and TeraW ($1\,\mathrm{TW} = 10^{12}\,\mathrm{W}$) are commonly used.

To quantify energy, the kWh is also used. Note that 1 Wh is the energy delivered for 1 h by a system with a power of 1 W. As a result:

$$1\,\mathrm{Wh} = 3{,}600\,\mathrm{J}$$

Similarly, $1\,\mathrm{kWh} = 10^3\,\mathrm{Wh}$, $1\,\mathrm{MWh} = 10^6\,\mathrm{Wh}$, $1\,\mathrm{GWh} = 10^9\,\mathrm{Wh}$, $1\,\mathrm{TWh} = 10^{12}\,\mathrm{Wh}$.

To assess an energy system, it is necessary to know both its power and the energy delivered during a reference period. If, for example, we know the nominal power of a hydroelectric plant $\wp = 50$ MW and the energy it produces in one year, $E = 200$ GWh, these two pieces of information tell us that the installation has supplied the energy in one year as if it had operated at its rated speed for the duration:

$$T = \frac{E}{\wp} = 4{,}000\ \mathrm{h}$$

The time unit of the result is in seconds if the energy and the power are given in J and in W. It is in hours if the energy and the power are given in Wh and in W.

A year has 8,760 h. A duration of operation at rated power of 4,000 h means that the installation has operated at the rated power for 46% of the duration of one year. We use the load factor:

$$f = \frac{E}{\wp T_{year}}$$

to assess the use of an installation. T_{year} is the duration of a year, expressed in the unit corresponding to the units used to express power and energy. It is easily understood that it is economically more advantageous to design an installation of lower power if the load factor is too small due to a limited energy resource. It is possible to make such a choice for a hydroelectric installation. On the other hand, for wind energy for example, the load factor is fixed by the wind conditions, which are random by nature.

The powers and energies vary considerably depending on the dimensions of the systems considered. It is essential to have a few orders of magnitude in mind to judge the relevance of a dimensioning.

Power	
Light bulb for lighting a living room	10–100 W
Champion cyclist	400 W
Appliances (washing machine, dishwasher)	1 kW
Maximum electrical power supplied in a house	4–6 kW
Power supplied by an onshore wind turbine	1 MW
Power of a nuclear power plant unit	900–1,300 MW
Record electricity consumption in France	100,000 MW
Energy	
Electricity consumption of a family in France	10,000 kWh
Annual electricity consumption in France	500 TWh
Calorific value of a liter of gasoline	10 kWh
Fuel consumption of a combustion engine car per 100 km traveled	50–100 kWh
Consumption of an electric car per 100 km traveled[2]	25 kWh
Price including VAT of electricity in France for domestic consumption (including consumption and subscription)[3]	€0.19 per kWh, €190 per MWh

Table I.1. *Orders of magnitude of powers and energies*

Table I.1 indicates orders of magnitude of power and energy. The data presented are essentially consumption data. In terms of production, only the power of an onshore wind turbine and the power of a nuclear power plant unit are shown in the table. More precise figures on production are provided in the chapters of this book for the different technologies studied. Table I.1 also indicates the order of magnitude of the price of domestic electricity. This is a price that includes distribution, taxes and VAT. It obviously differs from the electricity negotiation prices between producers and distributors, which vary according to supply and demand.

2 We should not conclude from these figures a definitive advantage of the electric car. This depends on the mode of production of the electricity that powers the electric vehicle. The efficiency of an internal combustion engine varies between approximately 30 and 50%.

3 Eurostat data for the year 2020.

I.3. The sources of this book

The content of this book was developed as part of the teaching provided by its author within the international master's degree in Simulation and Optimization of Energy Systems (SIMOS) awarded at the École nationale supérieure en génie des technologies industrielles, an engineering school belonging to the University of Pau and Pays de l'Adour.

This master's degree emphasizing energy optimization, we find in many exercises of this book an optimization point of view for dimensioning equipment.

The pedagogical approach of this book is built around the exercises and their resolution. The author is aware that the theoretical bases presented in Chapters 1 and 2 will be indigestible for some readers. However, he did not want to develop them any further. This book is not a textbook. It suggests that the reader approaches the resolution of the exercises directly in order to come gradually to the theory and master its implementation by going back and forth between the exercises and the theoretical parts.

I.4. Energy conversion

Energy comes in several forms: electrical energy, thermal energy, mechanical energy, chemical energy, etc. Converting energy means changing one form of energy into another. The emphasis in this book is on mechanical conversion. Different forms of mechanical energy appear in this work. The mechanical energy of a rotating solid, such as a rotor, is kinetic energy. The mechanical energy of a fluid is quantified by its head:

$$H = p + \rho\frac{\vec{u}.\vec{u}}{2} + \rho g z$$

where the pressure p, the kinetic energy (depending on the speed $\vec{u}$ and the density ρ) and the potential energy ($g = 9.81\ \mathrm{m.s^{-2}}$ is the acceleration of gravity and z designates the altitude) intervene. The head is usually evaluated in units of pressure (the Pascal, Pa). It is important to remember that head is energy per unit volume ($1\ \mathrm{Pa} = 1\ \mathrm{J/m^3}$).

In the exercises, we will see many types of mechanical conversion between these different forms of energy.

The power associated with a force $\vec{F}$ applied to a solid which makes it move at the speed $\vec{u}$ is:

$$\wp = \vec{F}.\vec{u}$$

Similarly, the power associated with a rotor to which a torque C is applied and which rotates at angular velocity ω is:

$$\wp = C.\omega$$

Finally, in a similar way, we consider a volume V containing a fluid. In steady state, for an incompressible fluid and an isothermal flow, the volume flow rate Q entering the volume must be compensated by the same flow rate leaving the volume. The power required to maintain this flow is the flow rate multiplied by the head variation between the volume inlet and outlet:

$$\wp = \Delta H.Q$$

The object is not to go into the detail of these equations here but to underline their similarities. For a point solid, power is a trade-off between force and speed. There is no power without speed or force. For a rotating rotor, power is a compromise between rotational speed and torque. For a fluid in permanent flow, the power is a compromise between flow rate and head variation.

This notion of compromise is naturally felt by a cyclist pedaling on their bike. To provide power to the extent of their abilities, they will choose the appropriate pinion to feel comfortable in terms of rotational speed and torque to be applied to the pedals.

If there is a common link between all the problems dealt with in this book, it is that of finding the appropriate compromise between force and speed, torque and speed of rotation, head and flow rate. All the exercises in this book ultimately deal with this same question.

1

Revision of Fluid Mechanics

This chapter summarizes the knowledge in fluid mechanics necessary for the resolution of the exercises. It is a tool to help the reader avoid having to look for equations in fluid mechanics books. The reader should refer to these works[1] to read the mathematical proofs that are not reproduced here. This chapter presents the scientific approach and identifies the specificities of the treated flows.

1.1. Euler's equations

Euler's equations are the equations whose solution is the velocity field $\vec{u}(t,\vec{x})$ and the pressure field $p(t,\vec{x})$ in a fluid domain for a fluid of zero viscosity. From the velocity field, we determine the trajectory $\vec{x}(t)$ of a fluid particle, whose position is at the point M_0 at the moment $t = 0$ (vector $\vec{x}_0$). At the moment t, the fluid particle is at point M and the positions and velocities are mathematically described by the following equations:

$$\overrightarrow{OM}(t,\vec{x}_0) = \vec{x}(t)$$

$$\vec{u}(t,\vec{x}(t)) = \frac{d\vec{x}}{dt}$$

A current line is schematized in Figure 1.1. When a flow is permanent, the velocity field and the pressure field are independent of time. Streamlines and trajectories are identical. It is then interesting to describe the flow in the intrinsic coordinate frame, linked to the trajectory of the particles. The displacement of a

1 A previous work by the author is in the spirit of this work: Mory M. (2011). *Fluid Mechanics for Chemical Engineering*. ISTE Ltd, and John Wiley and Sons, New York. The old book by Landau and Lifshitz (1959) is timeless.

fluid particle can be measured by its curvilinear abscissa $s(t)$ along the trajectory; it is the distance traveled on this line from an origin position M_0.

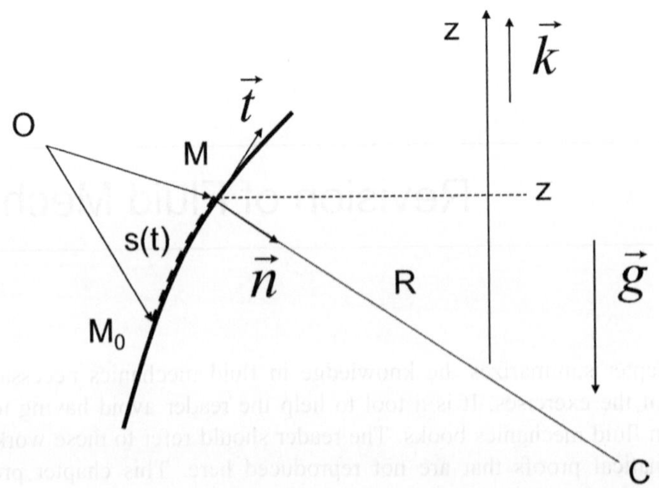

Figure 1.1. *Definition of the intrinsic coordinate system*

To describe the movement in the intrinsic coordinate system, the tangent vector $\vec{t}$ and the vector $\vec{n}$ normal to the trajectory are used. Two results of differential geometry are useful.

The tangent vector is calculated by the expression:

$$\vec{t} = \frac{d\overrightarrow{OM}}{ds}$$

The vector normal to the trajectory line, directed toward the interior of the curve, is such that:

$$\frac{d\vec{t}}{ds} = \frac{\vec{n}}{R}$$

R designates the radius of curvature of the trajectory. In Figure 1.1, point C is the center of curvature of the trajectory. This figure also indicates the acceleration of gravity $\vec{g}$, along the direction opposite the unit vector $\vec{k}$, and the altitude z of point M.

The velocity at the point M, collinear with the unit vector $\vec{t}$, is written as:

$$\vec{V} = V\vec{t} = \frac{ds}{dt}\vec{t}$$

Euler's equations are then written, for a permanent flow in the plane $(\vec{t},\vec{n})$:

– according to the unit vector $\vec{t}$:

$$\rho V\frac{\partial V}{\partial s} = -\frac{\partial p}{\partial s} + \rho(\vec{g}.\vec{t})$$

– according to the unit vector $\vec{n}$:

$$\rho\frac{V^2}{R} = -\frac{\partial p}{\partial n} + \rho(\vec{g}.\vec{n})$$

Euler's equations are the Navier–Stokes equations for a fluid whose viscosity is zero. By definition, an *ideal fluid* is a fluid of zero viscosity. We avoid using the term *ideal fluid* in this work because, we will explain later why, the application of this term is ambiguous. It is not a property of the fluid if we can use Euler's equations but a property of the flow.

We use Euler's equations under conditions where the flow is steady, the fluid is incompressible and the flow is isothermal. The density ρ is therefore taken as a constant.

Some additional calculations, from the spring of the differential geometry, allow transforming the terms of gravity and writing the Euler's equations:

$$\rho V\frac{\partial V}{\partial s} = -\frac{\partial p}{\partial s} - \frac{\partial(\rho g z)}{\partial s}$$

$$\rho\frac{V^2}{R} = -\frac{\partial p}{\partial n} - \frac{\partial(\rho g z)}{\partial n}$$

We then finally write:

– according to the unit vector $\vec{t}$:

$$\frac{\partial}{\partial s}\left\{p + \rho g z + \rho\frac{V^2}{2}\right\} = 0$$

Under the assumption that the quantity inside the parenthesis is a continuously differentiable function along the streamline, Bernoulli's theorem (see section 1.2) follows from this first equation:

– according to the unit vector $\vec{n}$:

$$\rho\frac{V^2}{R} = -\frac{\partial}{\partial n}\{p + \rho g z\}$$

The consequences of this second equation are discussed in section 1.3.

1.2. Head and Bernoulli's theorem

Euler's equation according to the tangent vector $\vec{t}$ leads to introduce the head, defined as the quantity:

$$H = p + \rho g z + \rho\frac{V^2}{2}$$

Bernoulli's theorem is deduced by integrating Euler's equation according to the tangent vector $\vec{t}$: *the head is conserved along a streamline.*

This very simple proof of Bernoulli's theorem assumes *incompressible fluid, steady* and *isothermal* flow, and *zero viscosity*. It also makes the assumption that the head is a continuously differentiable function along the streamline. Nothing states that this is always the case.

The problem with the term *ideal fluid* is that the application of Bernoulli's theorem is then prohibited for common fluids such as water and air, since their viscosity is not zero, although low. Bernoulli's theorem is however very useful, in water as in air, and it would be very regrettable to refrain from using it for scientific rigor. Experience shows that the head can be retained on certain portions of a streamline without being on other portions of the same streamline. It is therefore a property of the flow when the head is conserved or not. It is not a property of the fluid. The notion of ideal fluid does not explain why the head is kept along a streamline portion and not along another portion of this same streamline.

To explain this point with an example, we discuss the emptying flow of a tank filled with water, schematized in Figure 1.2. The upper reservoir is filled with water to a height H above an orifice (point B) through which the fluid flows with a flow rate velocity U_B. The flow in the upper reservoir, with the jet exiting it, is called the Torricelli flow. We extend this flow by an equally classic configuration where the jet of water issuing from point B of the upper reservoir flows into the air before plunging into a lower reservoir where, sooner or later, any fluid particle coming from the upper container ends up having zero velocity. We therefore consider the flow along a streamline between a point A placed on the surface of the water in the upper container and a point D on the surface of the water in the lower container. The pressure is the atmospheric pressure in A and in D since we are on the surface of the water in contact with the air. The value of the head is easily calculated at A and at D, since the kinetic energy is negligible at these points (when the tanks are large), and it can be seen that:

$$H_A = P_{atm} + \rho g(H + h) > H_D = P_{atm}$$

The head is therefore not conserved on the current line which leads from point A to point D. However, the first use made classically of Bernoulli's theorem consists of applying it between points A and B to determine the velocity of the jet U_B at B and deduce Torricelli's formula from it $U_B = \sqrt{2gH}$. It is therefore downstream of point B that the head loss occurs.

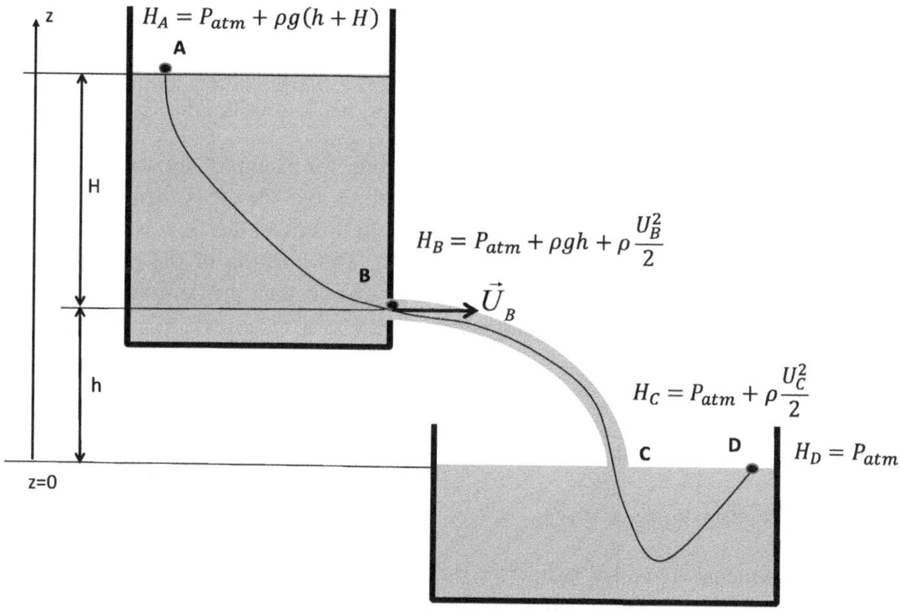

Figure 1.2. *Drainage flow from a tank*

Before going further, note that the head has an energetic significance since its dimension, the Pascal, is also understood as a Joule per unit volume. The head is the sum of three mechanical energies per unit volume: pressure, potential energy and kinetic energy. A flow (in steady state) is therefore characterized by transfers between these three forms of energy. By considering the values of the head indicated in Figure 1.2 at points A, B, C and D, we can therefore see that:

– The kinetic energy increases between A and B, compensated by a drop in potential energy. The same is true between B and C.

– On the other hand, between C and D, the kinetic energy must decrease. If we want the head to be conserved between C and D, this could only be done by a rise in pressure or potential energy. This is not what experience shows. To do this, it would be necessary to imagine a jet which springs from the lower tank to go and fill another upper tank.

Generally speaking, Bernoulli's theorem supports the idea that transfers between pressure, potential energy and kinetic energy are reversible. This is a misconception. In reality, it is easy to convert pressure energy or potential energy into kinetic energy. Accelerating a fluid by taking advantage of potential energy or pressure usually does well. It is therefore common to apply Bernoulli's theorem between two points between which the fluid is accelerated. Conversely, when it is necessary to reduce the velocity of the fluid between two points, the transfer of kinetic energy into pressure or potential energy is less simple. When the requested deceleration is too fast, a head loss often occurs.

Some flows considered in this work (inside a rotor for example) are stationary in a frame rotating at the angular velocity ω around an axis Oz. It is then useful to express Bernoulli's theorem in a rotational frame. In this case, the flow is described in a frame of cylindrical coordinates (Figure 1.3). The position of the point M is determined by the coordinates (r, θ, z) and, taking into account the definitions of the radial $\vec{e}_r$, azimuthal $\vec{e}_\theta$ and axial $\vec{e}_z$ unit vectors:

$$\overrightarrow{OM} = r\vec{e}_r + z\vec{e}_z$$

The velocity is written in the rotating cylindrical coordinate system as:

$$\vec{u}' = u'_r\vec{e}_r + u'_\theta\vec{e}_\theta + u'_z\vec{e}_z$$

The components are called radial (u'_r), azimuthal (u'_θ) – it is the component of rotation around the axis Oz – and axial (u'_z) component. The velocity in the fixed frame of cylindrical coordinates of the same axis Oz is by the principle of the composition of velocities[2]:

$$\vec{u} = \vec{u}' + \omega r\vec{e}_\theta$$

2 By convention in this work, the velocity in the fixed frame is denoted as $\vec{u}$ while the velocity in the moving frame is denoted as $\vec{u}'$.

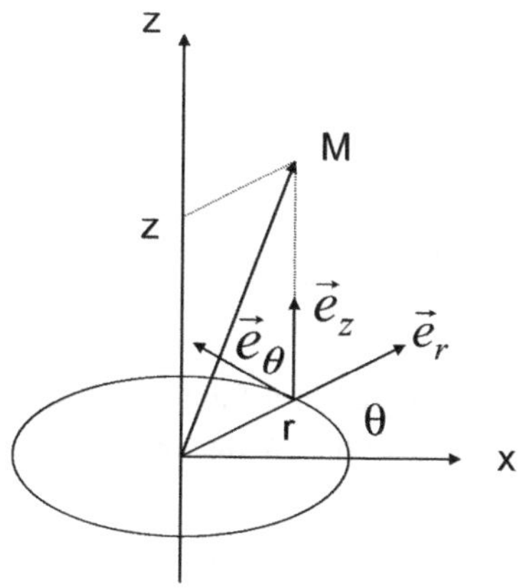

Figure 1.3. *Cylindrical coordinate system*

Euler's equations in the rotating cylindrical coordinate system are written as:

$$\rho\left\{u'_r\frac{\partial u'_r}{\partial r}+\frac{u'_\theta}{r}\frac{\partial u'_r}{\partial \theta}+u'_z\frac{\partial u'_r}{\partial z}-\boldsymbol{\frac{u'^2_\theta}{r}}-\boldsymbol{2\omega u'_\theta}\right\}=-\frac{\partial}{\partial r}\left(p-\rho\omega^2\frac{r^2}{2}\right)$$

$$\rho\left\{u'_r\frac{\partial u'_\theta}{\partial r}+\frac{u'_\theta}{r}\frac{\partial u'_\theta}{\partial \theta}+u'_z\frac{\partial u'_\theta}{\partial z}+\boldsymbol{\frac{u'_r u'_\theta}{r}}+\boldsymbol{2\omega u'_r}\right\}$$
$$=-\frac{1}{r}\frac{\partial}{\partial \theta}\left(p-\rho\omega^2\frac{r^2}{2}\right)$$

$$\rho\left\{u'_r\frac{\partial u'_z}{\partial r}+\frac{u'_\theta}{r}\frac{\partial u'_z}{\partial \theta}+u'_z\frac{\partial u'_z}{\partial z}\right\}=-\frac{\partial}{\partial z}\left(p-\rho\omega^2\frac{r^2}{2}\right)$$

Terms (in bold) with no partial derivative in the left-hand side of equality in the equations according to $\vec{e}_r$ and $\vec{e}_\theta$ are called centrifugal force and Coriolis force. In these equations, the pressure p designates the piezometric pressure (section 1.3).

By multiplying the three equations by u'_r, u'_θ and u'_z, respectively, then by summing the three equations obtained, we deduce:

$$\rho\left\{u'_r\frac{\partial}{\partial r}+\frac{u'_\theta}{r}\frac{\partial}{\partial \theta}+u'_z\frac{\partial}{\partial z}\right\}\left(\frac{u'^2_r}{2}+\frac{u'^2_\theta}{2}+\frac{u'^2_z}{2}\right)$$
$$=-\left\{u'_r\frac{\partial}{\partial r}+\frac{u'_\theta}{r}\frac{\partial}{\partial \theta}+u'_z\frac{\partial}{\partial z}\right\}\left(p-\rho\omega^2\frac{r^2}{2}\right)$$

The remarkable property of this result is that the centrifugal and Coriolis terms have disappeared to the left of the equality. By following a streamline along which the curvilinear abscissa is noted $s(t)$, the previous equation is reformulated in the form:

$$\rho\frac{ds}{dt}\frac{d}{ds}\left(p-\rho\omega^2\frac{r^2}{2}+\frac{u'^2_r}{2}+\frac{u'^2_\theta}{2}+\frac{u'^2_z}{2}\right)=0$$

Bernoulli's theorem when applied in a frame rotating at the angular velocity of rotation ω around an axis Oz is therefore written as:

$$p+\frac{u'^2_r}{2}+\frac{u'^2_\theta}{2}+\frac{u'^2_z}{2}-\rho\omega^2\frac{r^2}{2}=\text{constant}$$

along the current line. p refers here to the piezometric pressure. The new term is the last term to the left of the equality. It is a term of centrifugal force due to the fact that the frame rotates. When Bernoulli's theorem is applied in a rotating frame, it usually results in an increase in pressure, as we move away from the axis of rotation, due to centrifugal force[3].

1.3. Hydrostatics and variation of pressure in a direction perpendicular to a streamline

Euler's equation according to the vector normal $\vec{n}$ to the trajectory:

$$\rho\frac{V^2}{R}=-\{p+\rho gz\}$$

provides information on the pressure variation in the plane normal to the direction of the flow.

When the radius of curvature R of the streamline tends to infinity, the preceding equation indicates that $p+\rho gz=$ constant in the plane normal to the direction of the flow. In this case, the flow is said to be a parallel flow; the pressure is then hydrostatic in the plane perpendicular to the direction of flow.

The quantity $p+\rho gz$ is called piezometric pressure (or driving pressure). It is often used instead of pressure without giving it a specific notation. The advantage of using the piezometric pressure in the equations of fluid mechanics (when the density of the fluid is constant and homogeneous) is twofold. The first interest is that it

3 These properties are analyzed in the exercise proposed in section 2.4.

avoids dragging the gravity terms into the equations and makes it possible to identify the pressure variations due to the flow independently of those due to the change in altitude of the fluid particle. Second interest, resulting from the previous principle: it is often useful to be able to consider the piezometric pressure be constant on a surface through which the flow is parallel (see section 1.8).

Conversely, when the radius of curvature R is finite, since $\vec{n}$ is the normal vector directed toward the interior of the trajectory, the piezometric pressure $p + \rho g z$ increases in the plane perpendicular to the direction of the flow when one moves away from the center of curvature.

1.4. Linear head losses

Linear head losses are the energy losses caused by viscous friction on the walls of a cylindrical pipe. We consider a pipe of length L and diameter D in which flows a fluid of density ρ and kinematic viscosity ν. The pipe being cylindrical, its diameter is constant, its flowing section $S = \pi D^2/4$ is also constant, just like the flow velocity $U = Q/S$. The quantity Q designates the flow rate. The flow velocity depends on the variation of the piezometric pressure $\Delta P = \Delta H$ (which is equal to the head variation since the velocity does not change) between the entry and the exit. Flow is directed from the high-pressure inlet to the low-pressure outlet.

The relation between the head variation ΔH and the discharge velocity U is described by a relation between three dimensionless numbers:

– the linear head loss coefficient:

$$\lambda = \frac{2\Delta P D}{\rho U^2 L}$$

– the Reynolds number:

$$Re = \frac{UD}{\nu}$$

– the relative roughness:

$$k = \frac{e}{D}$$

Roughness e, which has the dimension of a length, is specific to the material constituting a pipe and its condition. Reference values are given in Table 1.1.

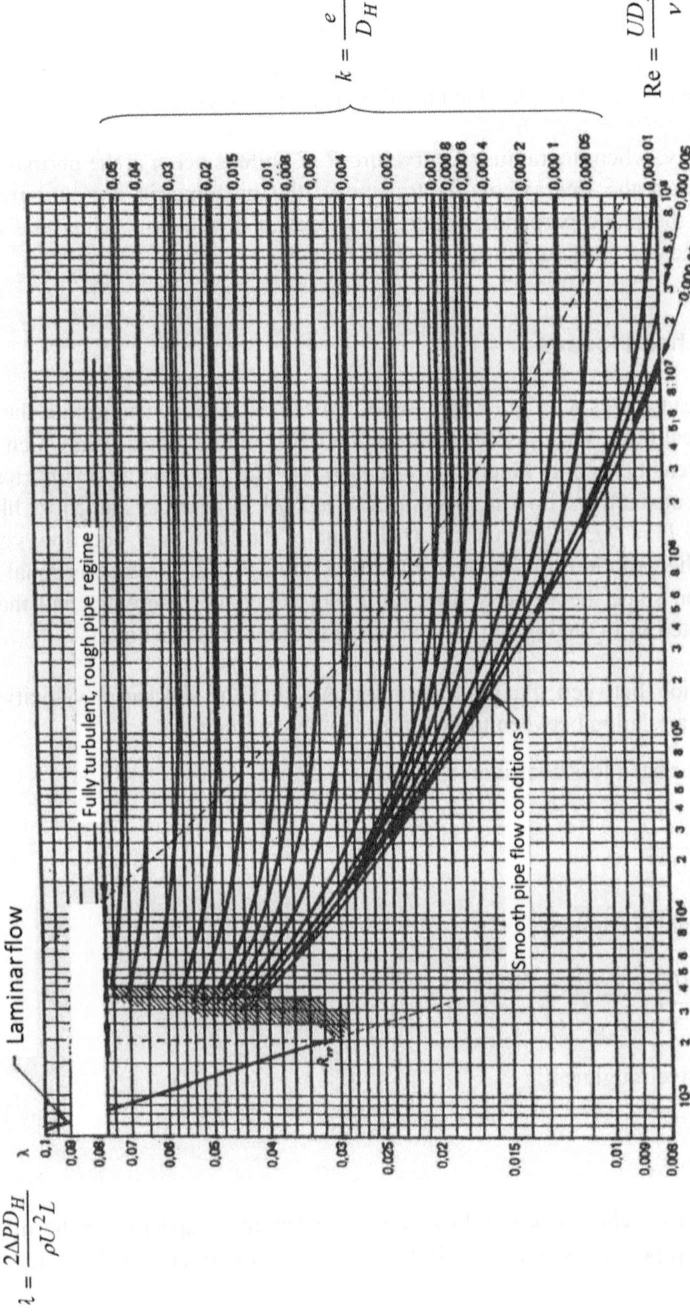

Figure 1.4. *Colebrook diagram (also called Moody diagram) for linear head losses*

Material	Conditions	e (in mm)
Brass, copper, aluminum, plastics, glass	Smooth, no deposits	e < 0.03
Steel	New Slightly rusty Rusty Calcified Asphalted, new Asphalted, normal Galvanized	0.03 < e < 0.1 0.1 < e < 0.2 0.2 < e < 0.3 0.5 < e < 2 0.03 < e < 0.05 0.1 < e < 0.2 e = 0.13
Cast iron	New Rusty Calcified Asphalted, new	e = 0.25 1 < e < 1.5 e > 1.5 0.03 < e < 0.05
Concrete	–	0.3 < e < 3

Table 1.1. *Absolute pipe roughnesses*

Colebrook diagram (Figure 1.4) shows the relationship $\lambda(Re, k)$.

In the formulas of Figure 1.4, the diameter D_H designates the hydraulic diameter linked to the ratio of the surface of the flow section S to its perimeter P:

$$D_H = \frac{4S}{P}$$

Colebrook diagram applies to the linear head loss in cylindrical pipes of any section. For a circular pipe or a pipe with a square section, we check that the hydraulic diameter is equal to the diameter or to the side of the square, respectively.

Three additional remarks are useful in the context of the applications dealt with in this work:

– In industrial-scale installations, the flow is almost always turbulent, with high values of Reynolds number.

– The values of the head loss coefficient vary little. They are most often included between $0.01 < \lambda < 0.03$. Taking linear head losses into account in a design is only important if the ratio L/D is large enough.

– For each value of the roughness coefficient k, the linear head loss coefficient is independent of Reynolds number beyond a certain value. In Figure 1.4, the *rough regime* conditions are plotted in space to the right of the dotted line. In this domain, the linear head loss coefficient only depends on the roughness; it does not vary with the flow velocity or with the viscosity of the fluid. We often have this case in industrial situations and this simplifies the resolution, subject to checking *in fine* that we are indeed in a rough regime (see the exercise proposed in section 5.2).

1.5. Singular head losses

Singular head losses are sudden head losses undergone by a flow, which are often produced by a sudden change in flow section. As explained in section 1.2, head losses occur most often when a fluid is decelerated and the flow would then have, in order to keep the head constant, to convert part of its kinetic energy into pressure energy or potential energy.

In the case of a sudden widening (Figure 1.5(a)), the head loss can be evaluated by Borda's formula:

$$\Delta H = \rho \frac{U_1^2}{2}\left(1 - \frac{S_1}{S_2}\right)^2 = \rho \frac{U_2^2}{2}\left(1 - \frac{S_2}{S_1}\right)^2$$

This formula involves the flow sections S_1 and S_2, as well as the velocities U_1 and U_2 in those respective sections, bound by conservation of flow rate $U_1 S_1 = U_2 S_2$.

In Figure 1.5(b), the configuration of a sudden narrowing is schematized. This diagram does not contradict the principle that the loss of head occurs during the deceleration of the fluid. In the flow diagram drawn, a separation occurs at the edge of the constricted section and the flow is accelerated until the section S_c. There is no head loss upstream of S_c in the volume in light gray. The head loss occurs in the dark gray area (as in Figure 1.5(a)) where the flow is decelerated between the section S_c and the section S_1.

The singular head losses are always evaluated by an equation of the form:

$$\Delta H = \zeta \rho \frac{U^2}{2}$$

A reference velocity U is chosen for the flow (most often a flowing velocity in a given section) and the coefficient of singular head loss ζ, without dimension, evaluates the head loss produced in relation to the kinetic energy. When using values of the head loss coefficient taken from the literature[4], care must be taken to take the reference velocity that corresponds to it. For Borda's formula, for example,

4 For example, Idelchik, I.E. (1993). *Handbook of Hydraulic Resistance*. Begell House, New York.

the discharge velocity U_1 in the narrow section is generally chosen and $\zeta_1 = (1 - S_1/S_2)^2$. We could just as naturally choose the output velocity U_2 in the widened section, in which case $\zeta_2 = (1 - S_2/S_1)^2$. With flow rate conservation $U_1 S_1 = U_2 S_2$, the two expressions give the same value of the head loss.

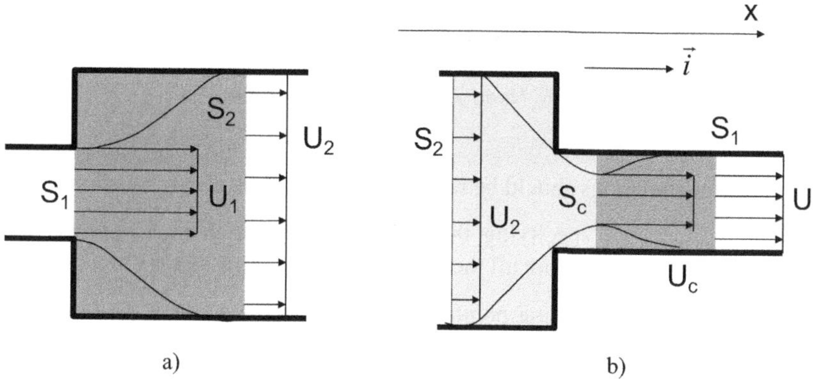

Figure 1.5. *Sudden enlargement and sudden contraction*

The singular head loss coefficients ζ generally depend only on the geometric characteristics of the components which cause them. They do not depend on the viscosity of the fluid, in the same way that the coefficient of linear head loss in rough regime does not depend on Reynolds number (section 1.4).

For Borda's formula, for example, the singular head loss coefficient $\zeta_1 = (1 - S_1/S_2)^2$, for the reference velocity U_1, depends only on the ratio of sections.

1.6. Head variation along a streamline

Along a streamline (Figure 1.6), the change in head between an entry point E and an exit point S is governed by the following equation (which some call the generalized Bernoulli equation):

$$H_S - H_E = \sum_i \Delta H_{pump,i} - \sum_j \Delta H_{turbine,j} - \sum_m \lambda_m \rho \frac{U_m^2}{2} \frac{L_m}{D_{H,m}} - \sum_n \zeta_n \rho \frac{U_n^2}{2}$$

We recognize successively on the right of the equality the contributions of the pumps, the turbines, the linear head losses and the singular head losses. The head losses are identified with their respective velocities, head loss coefficients and geometric dimensions. Pumps bring energy to the fluid so that $\Delta H_{pump,i} > 0$, while turbines recover energy from the fluid (the term is negative for $\Delta H_{turbine,j} > 0$).

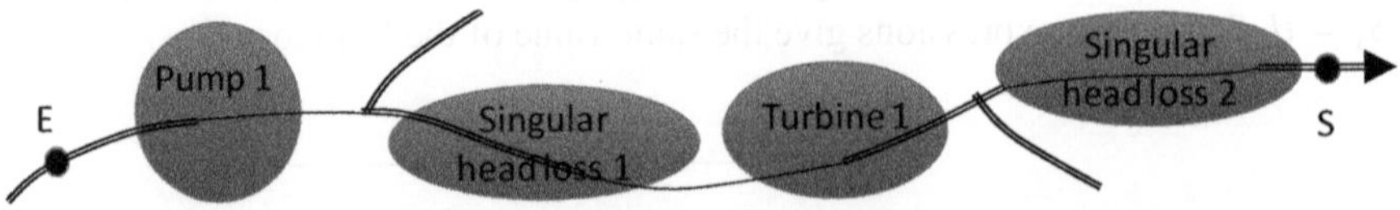

Figure 1.6. *Head variation along a streamline*

Two recommendations should be remembered to use the above equation wisely:

– the importance of specifying the streamline on which it is applied and of following it to take into account all the head losses that occur there;

– the usefulness of choosing points E and S in positions where the head can be determined (pressure, potential energy and kinetic energy).

The implementation of the generalized Bernoulli equation becomes all the more complex as one wishes to describe the flow in all its details. In this manual of exercises for the dimensioning of mechanical conversion systems (pumps, turbines, energy installations), our objective is to put our finger on the key points of dimensioning. In this spirit, only the head losses essential to the operation of the process will be taken into account. In a hydroelectric plant, for example, the linear head losses of the flows in the penstocks cannot be ignored due to the length of the pipes. On the other hand, the head loss due to a bend in the pipe can be considered secondary. Another example, if a stator is not installed downstream of a rotor to convert the rotational kinetic energy generated by the rotor into pressure, this rotational kinetic energy will undoubtedly be converted into a head loss which cannot be neglected. The spirit of these exercises is to size the system in an ideal configuration that allows understanding its operation, without masking this operation by parasitic phenomena.

1.7. Kinetic energy balance on a fluid volume

A complementary way of considering the conservation of mechanical energy consists of taking stock of it in a fluid domain V bounded by a surface S. The kinetic energy of the fluid contained in the volume is:

$$\iiint_V \frac{1}{2}\rho \vec{u}^2 \, dv$$

The temporal evolution of this quantity is written as:

$$\frac{\partial}{\partial t}\iiint_V \frac{1}{2}\rho(\vec{u}.\vec{u})dv = -\iint_S H\,(\vec{u}.\vec{n})ds - \iiint_V \rho\varepsilon dv$$

We are not giving the demonstration of this balance formula, which is a bit tedious, but classical[5]. The domain V and the surface S surrounding it are shown in Figure 1.8. The vector designates by convention the normal unit vector $\vec{n}$ at the surface S, directed toward the outside of the domain.

The head is defined by:

$$H = p + \rho g z + \frac{1}{2}\,\rho(\vec{u}.\vec{u})$$

For a Newtonian fluid, the rate of kinetic energy dissipation per unit mass is expressed:

$$\varepsilon = \frac{\nu}{2}\left\{\frac{\partial u_i}{\partial x_j} + \frac{\partial u_j}{\partial x_i}\right\}^2$$

where ν is the kinematic viscosity of the fluid. In this expression, the indices i and j are summed for i and j ranging from 1 to 3 using Einstein's notation[6]. The rate of kinetic energy dissipation quantifies the conversion of mechanical energy to thermal energy in a viscous fluid, which is found to be governed by velocity gradients in the fluid. The square in the expression for the rate of kinetic energy dissipation should be noted. The rate of kinetic energy dissipation is always positive, and velocity gradients add up to increase it.

In permanent mode, the kinetic energy contained in the fluid is constant and the term of left of the balance is null:

$$0 = -\iint_S H\,(\vec{u}.\vec{n})ds - \iiint_V \rho\varepsilon dv$$

We recognize in the two terms to the right of the equality:

– the head flow through the surface S surrounding the domain, which quantifies the flux of energy brought into the volume;

---

5 See, for example, Mory (2011).

6 Einstein's notation can be understood as $x_1 = x$, $x_2 = y$, $x_3 = z$ and $u_1 = u_x$, $u_2 = u_y$, $u_3 = u_z$.

– the integral of the viscous dissipation in the domain, which evaluates the energy dissipated in the volume.

If the viscosity is zero, the head flow entering the domain must be equal to the head flow leaving the domain. We find here a more general form than Bernoulli's theorem. If particles enter the domain and mix within the domain, the output head will be the average of the input heads, weighted by the flow rate.

We can apprehend important notions on turbulence by simply considering turbulence as a random movement of zero mean, which is superimposed on a mean flow. It is possible to measure at a point an average velocity $\vec{\bar{u}}$ and a fluctuation of velocity, the variance of which is noted u_{rms}. The kinetic energy is therefore split into two parts, the kinetic energy of the mean flow to which is added the mean kinetic energy of the turbulence:

$$\frac{1}{2}\rho(\vec{\bar{u}}.\vec{\bar{u}}) + \frac{1}{2}\rho u_{rms}^2$$

The bar on the velocity symbol $\bar{u}$, like those on pressure $\bar{p}$ or head $\bar{H}$ symbols, indicates the average in the meaning of turbulence. The previously written kinetic energy balance is then modified in the form:

$$\frac{\partial}{\partial t}\iiint_V \left\{\frac{1}{2}\rho(\vec{\bar{u}}.\vec{\bar{u}}) + \frac{1}{2}\rho u_{rms}^2\right\} dv$$
$$= -\iint_S \left\{\bar{H} + \frac{1}{2}\rho u_{rms}^2\right\}(\vec{\bar{u}}.\vec{n})ds - \iiint_V \rho\varepsilon dv$$

where the head's mean flow is:

$$\bar{H} = \bar{p} + \rho g z + \frac{1}{2}\rho(\vec{\bar{u}}.\vec{\bar{u}})$$

These elementary notions of turbulence are easily illustrated by analyzing the energy balance for the flow of a river passing over a weir (Figure 1.7). The sill is a sudden lowering of the bottom of the river. If the flow of the river is constant, we can apprehend the concept of stationary turbulence by looking at the surface of the river for a certain period of time to observe the variation in level or the movement of objects on its surface. Upstream of the sill and in the upper part of the waterfall, the level of the river is stable, the surface is smooth and the fluid particles passing at a point all follow the same trajectory. The flow is stationary in the exact sense of the term. It does not vary over time and the turbulence is undetectable. Downstream of the sill, the situation is more complex. Random eddies are observed which indicate that turbulence has been generated. Nevertheless, at any point, it is possible to evaluate an average level of the surface if we observe it for a certain time. When the flow is kept constant over time, if we manage to follow objects floating on the surface, we can also evaluate an average transport velocity of the fluid downstream.

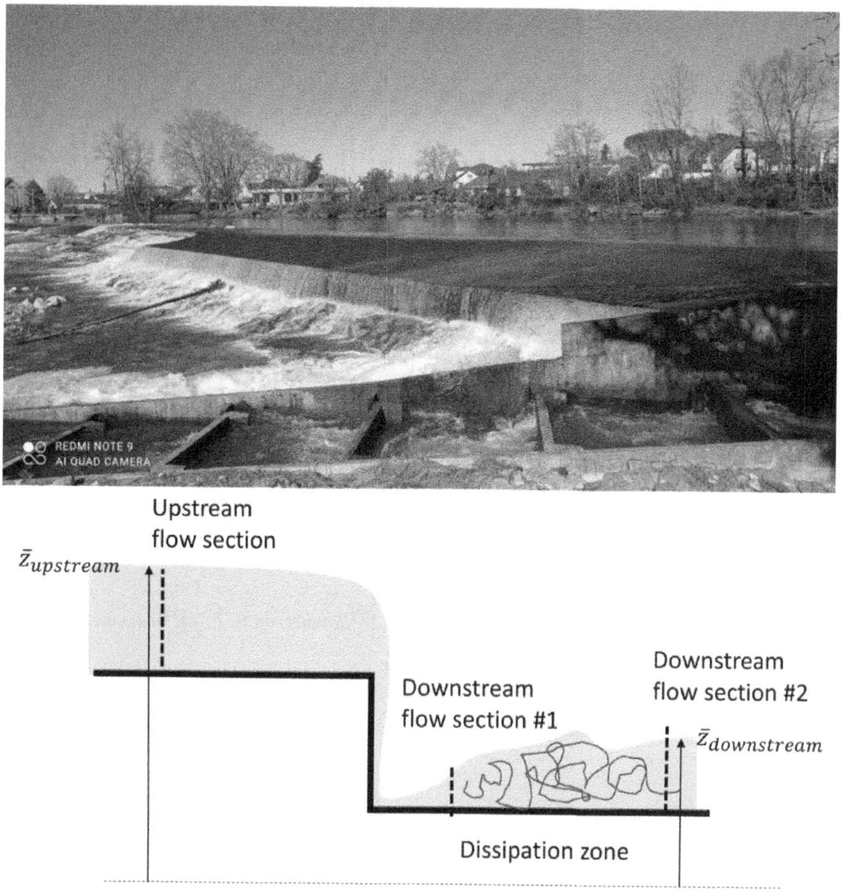

Figure 1.7. *Flow of a river over a sill: head loss, generation of turbulence and dissipation of kinetic energy. The photo shows the Heïd weir on the Gave de Pau in Bizanos (with fish pass in the foreground)*

Let us therefore now write the kinetic energy balance for a fluid volume between a flow section upstream of the sill and a flow section downstream of the sill in the swirl zone, as shown schematically on the drawing opposite the image shown in Figure 1.7. The situation being stationary in the mean sense of the turbulence, the balance is written as:

$$0 = -\left\{\overline{H}_{upstream} + \frac{1}{2}\rho u^2_{rms,upstream}\right\}\iint_{S_{upstream}} (\vec{u}.\vec{n})ds - \left\{\overline{H}_{downstream} + \frac{1}{2}\rho u^2_{rms,downstream}\right\}\iint_{S_{downstream}} (\vec{u}.\vec{n})ds - \iiint_V \rho\varepsilon dv$$

For simplicity, the heads and kinetic energies have been taken out of the integrals of the surface, assuming these quantities to be homogeneous in the upstream section and in the downstream section. By conservation of the flow rate Q between the two sections, we can write, taking into account the orientation of $\vec{n}$ (directed toward the outside of the domain V):

$$Q = -\iint_{S_{upstream}} (\vec{u}.\vec{n})ds = \iint_{S_{downstream}} (\vec{u}.\vec{n})ds$$

The balance sheet therefore simplifies to the form:

$$Q\left\{\overline{H}_{upstream} + \frac{1}{2}\rho u^2_{rms,upstream}\right\} - Q\left\{\overline{H}_{downstream} + \frac{1}{2}\rho u^2_{rms,downstream}\right\} = \iiint_V \rho\varepsilon dv$$

or again:

$$\overline{H}_{upstream} - \overline{H}_{downstream} = \frac{1}{2}\rho u^2_{rms,downstream} - \frac{1}{2}\rho u^2_{rms,upstream} + \frac{1}{Q}\iiint_V \rho\varepsilon dv$$

The term to the left of the equation is the mean flow head loss.

The flow of the river is a free surface flow. It is easy to analyze it by considering a streamline on the free surface (this streamline is called the water line) where the pressure is atmospheric pressure. Therefore:

$$\overline{H}_{upstream} - \overline{H}_{downstream} = \rho g \bar{z}_{upstream} - \rho g \bar{z}_{downstream} + \frac{1}{2}\rho \overline{U}^2_{upstream} - \frac{1}{2}\rho \overline{U}^2_{downstream}$$

Passing over a sill, the particles undergo a sudden head loss ($\overline{H}_{upstream} > \overline{H}_{downstream}$) due to the lowering of the water line ($\bar{z}_{upstream} > \bar{z}_{downstream}$). The fluid would have to be accelerated to compensate for it. This is what happens in the upper part of the waterfall. In the lower part of the waterfall, on the other hand, the acceleration of the fluid is no longer sufficient to compensate for the drop in potential energy due to the lowering of the water line. The potential energy is then partly converted into kinetic energy of the turbulent movements observed in the large eddies visible in the photo. The generation of turbulence therefore results from the head loss of the mean flow.

When the downstream section limiting the domain V is located at the beginning of the swirl zone (downstream flow section 1), the integral of the rate of dissipation of the kinetic energy on the domain is not sufficient to compensate for the head loss. The turbulent kinetic energy is then very important in the exit section. The dissipation rate varies with the square of the velocity gradients. At the beginning of the waterfall, it is low because the turbulent kinetic energy is also low. It is in the swirl zone that the dissipation of kinetic energy is most intense. The integral of the kinetic energy dissipation rate over the volume considered increases as the outlet section is moved downstream of flow section 2. The turbulent kinetic energy produced by the head loss has largely been dissipated and the turbulent kinetic energy at the outlet of the domain is lower.

The turbulent kinetic energy balance therefore makes it possible to understand two essential processes of turbulence:

– The turbulence is generated by head losses in the mean flow.

– Energy dissipation is accentuated in a turbulent flow.

1.8. Momentum theorem

Consider a fluid domain V (in gray) containing a fluid of density ρ as shown in Figure 1.8. It is surrounded by a closed surface $S = S_s \cup S_d$ on which we distinguish the surface S_s of the solid walls (solid lines) and the surface S_d (dashed lines), which is a virtual surface in the fluid through which the fluid can flow.

The momentum theorem relates the total force $\vec{F}$ applied by fluid flow on the solid surface S_s to the pressure force and momentum flux on the flow surface S_d. The first writing involves the piezometric pressure:

$$\vec{F} = -\iint_{S_d} \{(p + \rho g z)\vec{n} + \rho(\vec{u}.\vec{n})\vec{u}\}\, ds$$

and is the most common. It is easily implemented when the flow is parallel on the flowing surfaces S_d. The piezometric pressure is in this case constant on these surfaces (see section 1.3) and the pressure integrals are easily determined. The vector $\vec{n}$ is the normal unit vector at any point on the surface S_d by convention directed outward from the domain V. Figure 1.8 shows the case where the domain V contains a solid volume (in white). The wall of this volume must also be counted as part of the solid envelope and on this surface the vector $\vec{n}$ is also directed toward the exterior of the domain V.

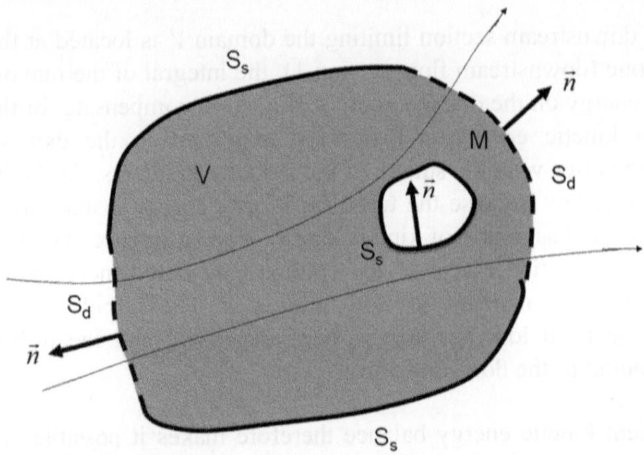

Figure 1.8. *Momentum theorem. Definition of the fluid volume V and the surfaces surrounding it*

The second expression of the momentum theorem separates the pressure from the force of gravity:

$$\vec{F} = -\iint_{S_d} \{p\vec{n} + \rho(\vec{u}.\vec{n})\vec{u}\}\, ds - \iiint_V \rho\vec{g}\, dv$$

It is less easy to use because the integral of the pressure force on the flowing surfaces must be calculated by taking into account the pressure variations due to gravity. It is nevertheless necessary to resort to it when the fluid filling the area V is a liquid and when part of the surface S_d is in contact with air. The pressure on this surface is then the atmospheric pressure; the first formulation does not allow calculating the piezometric pressure without knowing the variations of the altitude z of the surface in contact with the air.

The momentum theorem applies under the one and only condition that the *flow is permanent*. It does not assume any condition on the mechanical properties of the fluid and it applies to both Newtonian and non-Newtonian fluids.

The momentum theorem has two classical applications:

– If the force applied by the flow on the solid wall is known, it enables the pressure variation between the fluid inlet section and the fluid outlet section to be calculated. The head loss of the flow in the volume is deduced therefrom. It is by this principle that Borda's formula is obtained to estimate the head loss in a sudden widening (section 1.5).

– If the pressure on the flowing sections S_d is known, it makes it possible to calculate the force $\vec{F}$ applied to the solid wall S_s. It is this use of the momentum theorem that we will make in the exercises of this book to calculate the force applied to the blades of a rotor or a stator. The momentum theorem therefore plays a central role in this book.

To be complete in terms of useful mathematical tools, we recall Ostrogradski's formula which links the integral of a function f on a surface S surrounding a volume V to the integral of the gradient of this function in the volume:

$$\iint_S f\vec{n}\,ds = \iiint_V \nabla f dv$$

Ostrogradski's formula makes it possible to establish that the integral of a constant pressure force, for example, equal to the atmospheric pressure, on a closed surface is zero:

$$\iint_S P_{atm}\vec{n}\,ds = \vec{0}$$

since the function $f(x,y,z) = P_{atm}$ in the entire volume has zero gradient.

1.9. Angular momentum theorem

We mathematically define the moment with respect to a point O of a vector applied to the point M as the vector product of the vector $\overrightarrow{OM}$ with this vector. When this vector is the product of the density with the velocity vector $\vec{u}$, we thus define the moment of the momentum in M:

$$\overrightarrow{OM} \times (\rho\vec{u}) = \begin{pmatrix} x \\ y \\ z \end{pmatrix} \times \begin{pmatrix} \rho u_x \\ \rho u_y \\ \rho u_z \end{pmatrix}$$

We define in an analogous way the moment of a force $\vec{F}$ applied to the point M:

$$\overrightarrow{OM} \times \vec{F} = \begin{pmatrix} x \\ y \\ z \end{pmatrix} \times \begin{pmatrix} F_x \\ F_y \\ F_z \end{pmatrix}$$

The concepts of moment have an immediate utility when one considers a volume having a symmetry of revolution around an axis Oz. We then generally describe the flow in a cylindrical coordinate system (section 1.2 and Figure 1.3) with the velocity $\vec{u} = u_r\vec{e}_r + u_\theta\vec{e}_\theta + u_z\vec{e}_z$ for which the three components are, respectively, called the radial (u_r), azimuthal (u_θ) and axial (u_z) components.

The moment of the momentum and the moment of the force are expressed, respectively, in a cylindrical coordinate system:

$$\overrightarrow{OM} \times (\rho\vec{u}) = \begin{pmatrix} r \\ 0 \\ z \end{pmatrix} \times \begin{pmatrix} \rho u_r \\ \rho u_\theta \\ \rho u_z \end{pmatrix} = \begin{pmatrix} -z\rho u_\theta \\ z\rho u_r - r\rho u_z \\ r\rho u_\theta \end{pmatrix}$$

$$\overrightarrow{OM} \times \vec{F} = \begin{pmatrix} r \\ 0 \\ z \end{pmatrix} \times \begin{pmatrix} F_r \\ F_\theta \\ F_z \end{pmatrix} = \begin{pmatrix} -zF_\theta \\ zF_r - rF_z \\ rF_\theta \end{pmatrix}$$

The components according to z of these two vectors, respectively $r\rho u_\theta$ and rF_θ, have a particular role for a flow rotating around the Oz axis in a revolution volume[7].

7 The normal $\vec{n} = n_r\vec{e}_r + n_\theta\vec{e}_\theta + n_z\vec{e}_z$ to the surface surrounding a volume of revolution is such that $n_\theta = 0$.

When the flow is permanent in a revolution volume V, we deduce from the momentum theorem that the component of the torque applied by the flow on solid surfaces in the direction Oz is:

$$C_z = \iint_{S_s} (rF_\theta)ds = -\iint_{S_d} \rho(r\rho u_\theta)(\vec{u}.\vec{n})ds$$

S_s denotes the totality of the solid surfaces surrounding the domain V and S_d denotes the totality of the surfaces surrounding the domain V through which the flow rate is non-zero. $\vec{n}$ is the unit vector perpendicular to these surfaces, oriented outward from the domain V.

The expression for torque means that the torque applied to the solid surface surrounding the fluid domain is equal to the difference between the angular momentum inflow and the angular momentum outflow. We anticipate Chapter 2 by saying that this formula makes it possible to calculate the torque applied to the stator of a turbomachine. This equation is the angular momentum theorem.

The situation is more complex when one wants to consider the application of the angular momentum theorem to a machine rotor since the flow is permanent in a frame which rotates around the Oz axis with the angular velocity ω of rotation of the rotor. It is then necessary to write the equations in the cylindrical reference frame in rotation, which makes Coriolis terms appear. We finally obtain:

$$C_z = \iint_{S_s} (rF_\theta)ds = -\iint_{S_d} \rho(r\rho u'_\theta)(\vec{u}'.\vec{n})ds - \iiint_V \rho 2\omega r u'_r dv$$

The velocities are indicated $\vec{u}'$ since it is the velocity in the rotating frame that is distinguished from the velocity $\vec{u} = \omega r\vec{e}_\theta + \vec{u}'$ in the fixed frame. This writing of the angular momentum theorem is less advantageous because it is no longer enough to know the flow of angular momentum on the flowing surfaces surrounding the volume to calculate the torque applied to the solid surfaces. It is also necessary to know the radial component to integrate the Coriolis force over the entire volume. We come back to this question in sections 2.1.3.5 and 2.1.5.

1.10. Irrotational flows, potential flows

Knowing the velocity field $\vec{u}(x, y, z, t)$ in a volume V, it is possible to determine the rotational of the velocity which is written in a Cartesian coordinate system or in a cylindrical coordinate system:

$$\overrightarrow{Rot}(\vec{u}) = \begin{pmatrix} \frac{\partial u_z}{\partial y} - \frac{\partial u_y}{\partial z} \\ \frac{\partial u_x}{\partial z} - \frac{\partial u_z}{\partial x} \\ \frac{\partial u_y}{\partial x} - \frac{\partial u_x}{\partial y} \end{pmatrix} \qquad \overrightarrow{Rot}(\vec{u}) = \begin{pmatrix} \frac{1}{r}\frac{\partial u_z}{\partial \theta} - \frac{\partial u_\theta}{\partial z} \\ \frac{\partial u_r}{\partial z} - \frac{\partial u_z}{\partial r} \\ \frac{1}{r}\frac{\partial (r u_\theta)}{\partial r} - \frac{1}{r}\frac{\partial u_r}{\partial \theta} \end{pmatrix}$$

A flow is irrotational at a point if $\overrightarrow{Rot}(\vec{u}) = \vec{0}$.

It is useful to know if a flow is irrotational because, for an incompressible fluid and zero viscosity, Kelvin's theorem states that if the flow is irrotational in a fluid particle, it retains this property when we follow the particle in its movement. In other words, an irrotational flow is conserved. This property allows considering specific methods to determine the flow. These methods apply more particularly in aerodynamics to calculate the flow in a turbomachine (Chapter 2) or to describe the waves propagating on the surface of the sea (Chapter 8).

Since $\overrightarrow{Rot}(\nabla\phi) = \vec{0}$ for any differentiable function $\phi(x, y, z, t)$, it is advantageous to determine the velocity potential function $\phi(x, y, z, t)$ such as:

$$\vec{u} = \begin{pmatrix} u_x \\ u_y \\ u_z \end{pmatrix} = \begin{pmatrix} \frac{\partial \phi}{\partial x} \\ \frac{\partial \phi}{\partial y} \\ \frac{\partial \phi}{\partial z} \end{pmatrix}$$

for an irrotational flow in a fluid volume V. The knowledge of a single scalar function $\phi(x, y, z, t)$ is then sufficient to determine the three velocity components.

For an incompressible fluid, we also have:

$$div(\vec{u}) = \Delta\phi = 0$$

The potential function $\phi(x, y, z, t)$ is therefore a harmonic function. Its solution can be determined, depending on the initial conditions and the boundary conditions of the domain, by well-known specific methods. We will come back to this in the following chapters, when the need arises.

2

Hydraulic Turbomachines

2.1. General information on turbomachinery

2.1.1. *Turbopumps and turbines*

The main element of a roto-dynamic machine is its rotor, a symmetrical circular part rotating around an axis.

The machine is called a turbine when it converts the mechanical energy of a fluid flow into electricity. In this case, the fluid head:

$$H = p + \rho g z + \rho \frac{U^2}{2}$$

is weaker at the output of the machine than it is at the input, since the fluid gives up its energy to the machine. When the diameters of the pipes at the inlet and at the outlet of the machine are equal, the velocity is identical at the inlet and at the outlet. As the altitude between the inlet and the outlet varies little, the drop in the head results in a head loss between the inlet and the outlet.

The machine is a pump when it supplies mechanical energy to the fluid. In this case, the fluid head at the outlet is greater than the fluid head at the inlet, which results in a pressure increase.

A roto-dynamic machine actually couples two machines, a hydraulic machine and an electric machine. In a pump, the electric rotor (motor) drives the hydraulic rotor which pushes the fluid. In a turbine, the fluid flow sets in rotation the hydraulic rotor which drives the rotor of an alternator to produce electricity. A roto-dynamic machine therefore couples two rotors, a hydraulic rotor and an electric rotor, linked by a rigid axis. When the shaft rotates at constant angular speed ω, the component C_z of the torque applied by the electric machine to the shaft is equal and opposite to the torque applied by the fluid to the hydraulic wheel, by the mechanical principle of action/reaction.

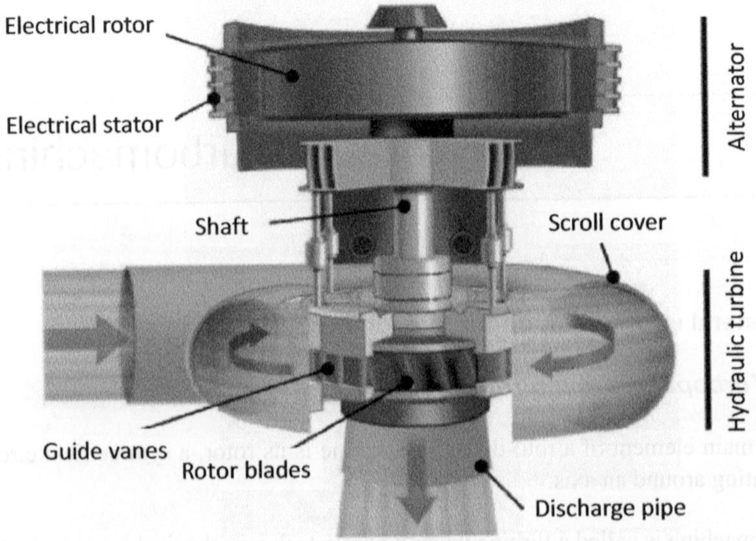

Figure 2.1. *Diagram of a Francis turbine[1]. For a color version of this figure, see www.iste.co.uk/mory/mechanical.zip*

The diagram of a Francis turbine (Figure 2.1) shows the coupling of the electric and hydraulic rotors through the shaft. The fluid flow enters through the spiral cover and exits parallel to the axis of rotation of the machine. We can roughly take the same diagram to represent a centrifugal pump, the essential difference being that the direction of the flow is reversed. It enters the machine through the suction pipe centered on the axis and exits through the spiral case, which is also called a scroll casing or volute.

When the machine is operating under ideal conditions, that is, when the conversion of energy between the fluid and the machine takes place without loss, the mechanical power supplied or recovered by the fluid is equal to the power electricity delivered or consumed, that is:

$$\wp_{elec} = \wp_{meca} = C_z.\omega = \Delta H_{machine}.Q$$

1 Available at: http://www.hydroquebec.com/comprendre/hydroelectricite/turbine-alternateur.html.

Q is the volume flow rate of the fluid passing through the machine and $\Delta H_{machine}$ is the head given up or recovered by the fluid, depending on whether the machine is a turbine or a pump. The second writing of the mechanical power involves the torque C_z and the angular speed of rotation ω.

2.1.2. *Notions of aerodynamics*

Understanding the operation of a rotor requires knowledge of a few notions of aerodynamics in order to introduce the concept of the lift force applied to an aircraft wing by a flow. Figure 2.2 shows an airplane wing in a uniform fluid flow of speed $\vec{V}$. The wing deflects the fluid which flows around it. When the angle of incidence α is 0, the flow is symmetrical with respect to the center line of the wing since the latter is symmetrical. The pressures are identical on both sides of the wing and the lift force $\vec{F}_L$ applied by the flow on the wing, perpendicular to the direction of the wind, is 0. On the other hand, when the angle of incidence α is positive, the fluid accelerates more to bypass the wing from above than to pass below. By virtue of Bernoulli's theorem, the pressure applied to the upper side of the wing (suction side or extrados) is less than the pressure applied to the underside of the wing (pressure side or intrados). The wing undergoes a positive lift force which pushes it upwards in the direction perpendicular to the direction of the wind.

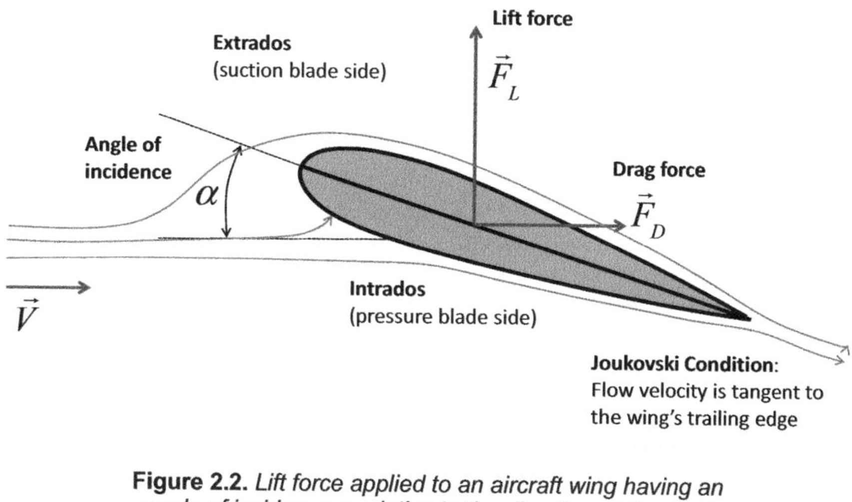

Figure 2.2. *Lift force applied to an aircraft wing having an angle of incidence α relative to the direction of the wind*

Figure 2.2 illustrates another characteristic of the flow. When all goes well, the streamlines bypassing the two faces of the wing remain attached to the wing and are both tangents to the trailing edge of the wing. This is called the Kutta–Joukovski condition.

The ideal operation described in Figure 2.2 is verified only for low angles of incidence. We define the lift coefficient by the dimensionless quantity:

$$C_z = \frac{F_L}{\rho \frac{V^2}{2} S}$$

where ρ is the density of the fluid, V the speed of the fluid upstream of the wing and S the area of the wing. Theory and experience show that the lift coefficient increases with the angle of attack (Figure 2.3) as long as the angle of attack is less than 10–15°. Beyond that, the lift drops. The phenomenon is called stall. Figure 2.3 shows the flow that occurs. The streamline passing over the extrados separates from the wing, because the fluid in the boundary layer does not have sufficient kinetic energy to increase the pressure. The Joukovski condition is no longer verified at the trailing edge of the wing and there is a boundary layer separation of the flow on the upper surface, which causes a head loss in the wake of the wing. This is obviously not a desirable situation since the lift is collapsing.

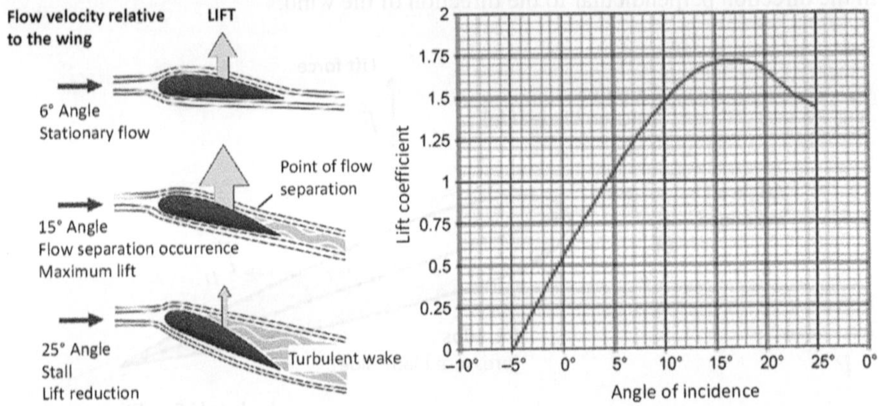

Figure 2.3. *Variation of the lift coefficient with the angle of incidence of the wing relative to the direction of the wind and illustration of the stall phenomenon*[2]

2 Available at: https://fr.wikipedia.org/wiki/Stall_(fluid dynamics) and https://fr.wikipedia.org/wiki/Lift_(force).

In what follows, we consider the rotor of a roto-dynamic machine as an assembly of blades in the shape of an aircraft wing. In a configuration where they work well, the Joukovski condition will apply to the trailing edge of each of them.

2.1.3. *Principle of mechanical conversion: force and power, velocity triangle, fixed reference and mobile reference*

2.1.3.1. *Force and power*

The example of an airplane wing discussed in the previous section shows that a flow around a solid element can apply a non-zero force $\overrightarrow{dF}_{fluid/solid}$ on this element. When the solid part moves at speed $\vec{u}_p$, this force generates a mechanical power which is written as:

$$d\wp_{meca} = \overrightarrow{dF}_{fluid/solid}.\vec{u}_p$$

If the speed is constant, the solid does not accelerate. It is then necessary, by virtue of the fundamental relation of mechanics, that the sum of the forces applied to the solid element be 0. This means that the electrical machine (motor or alternator) to which this mechanical element is linked must apply to it a force equal in absolute value and opposite in direction to the force $\overrightarrow{dF}_{fluid/solid}$. In other words, if we can neglect any loss (by mechanical friction in the bearings), the force $\overrightarrow{dF}_{fluid/solid}$ is entirely retransmitted to the electrical converter.

The first consequence of the expression of power is that a machine converts mechanical energy by its moving parts. The hydrodynamic rotor of a turbine converts the mechanical energy of the fluid into the mechanical energy of the rotating solid rotor. Then, the second stage of conversion, the alternator converts the mechanical energy of the solid rotor into electricity.

The second consequence is that the converted mechanical power is always a compromise between force $\overrightarrow{dF}_{fluid/solid}$ and speed $\vec{u}_p$. We presented this idea in the introduction, because it sums up in a certain way the argument of this work. All the exercises offered revolve around this trade-off between force and speed.

The first step in determining the converted mechanical power is to calculate the force applied to the solid walls of the rotor using the momentum theorem (section 1.8). This calculation is carried out in the exercises proposed in sections 3.3 (wind turbine rotor) and 4.3 (Kaplan turbine rotor) in a formulation where each element of the rotor is treated according to its radial position r.

2.1.3.2. *Velocity triangle*

The only condition required to apply the momentum theorem is that the flow is steady. This requires writing the momentum theorem in the frame linked to the blades, that is to say in the rotating frame to calculate the force applied to a rotor element. The speed $\vec{u}$ in the fixed reference and the speed $\vec{u}'$ in the reference of the blades are linked by the composition of the speeds:

$$\vec{u} = \vec{u}_p(r) + \vec{u}'$$

The dependence of the blade speed on the radius r is indicated to remind us that the blade speed varies between the hub and the tip of the blade. The velocities taken into account to apply the momentum theorem are the relative velocities $\vec{u}'$. After calculating the force $\overrightarrow{dF}_{fluid/solid}$, the converted power is obtained by multiplying the force by the speed of the blade $\vec{u}_p$.

The remarkable result of this calculation (see exercises in sections 3.3 and 4.3) is that the power $d\wp_{meca}$ converted by the rotor element at the radial position r depends only on three speed vectors, which specialists call the velocity triangle: the speed of the blade $\vec{u}_p(r)$, the speed of the fluid $\vec{u}'_{ups}$ upstream of the blade and the speed of the fluid $\vec{u}'_{downs}$ downstream of the blade. The converted power does not explicitly depend on the pressure difference between upstream and downstream.

2.1.3.3. *Head lost or recovered by the fluid*

The role of the pressure is identified when deducing the variation of head of the fluid particles in the machine from the power:

$$\Delta H_{machine} = \frac{d\wp_{meca}}{dQ}$$

dQ is the flow through the fluid volume surrounding this rotor element.

The head variation is also expressed as a function of the piezometric pressures and velocities upstream and downstream (taken in the fixed benchmark):

$$\Delta H_{machine} = \left| p_{ups} + \rho\frac{\vec{u}_{ups}.\vec{u}_{ups}}{2} - p_{downs} - \rho\frac{\vec{u}_{downs}.\vec{u}_{downs}}{2} \right|$$

In these expressions, the head variation and the power are positive values. The absolute value above makes it possible to write the same expression for a pump and for a turbine.

The piezometric pressure variation between upstream and downstream is therefore deduced from the head variation between upstream and downstream. We can say, to summarize, that the pressure variation between upstream and downstream serves to ensure the velocity triangle that governs the conversion process.

Let us complete this discussion with a last remarkable relation. If the head is calculated using the velocities in the mobile frame, after having determined $p_{ups} - p_{downs}$ from the head $\Delta H_{machine}$ using the speeds in the fixed frame, we obtain:

$$\Delta H'_{machine} = \left| p_{ups} + \rho \frac{\vec{u}'_{ups}.\vec{u}'_{ups}}{2} - p_{downs} - \rho \frac{\vec{u}'_{downs}.\vec{u}'_{downs}}{2} \right| = 0$$

This result[3] means that the energy conversion is ideal. The energy exchanged by the fluid is entirely recovered or released by the rotor, without causing any parasitic head loss.

2.1.3.4. *Fixed frame and moving frame*

The properties that we summarize in this section require the reader to be attentive to the frame (fixed or rotating) in which they apply the laws of mechanics.

To apply the momentum theorem, we use the velocities in the moving frame since the flow is stationary in this frame in the strict sense of the term. It is only in this frame that one can calculate the force applied by or on the rotor and the streamlines of the permanent flow. The blade is stationary and the head variation, calculated with the speeds in the moving frame, is zero if the energy conversion is perfect.

On the other hand, it is necessary to place oneself in the fixed frame (speed of the flow and speed of the blade) to calculate the converted mechanical power and link it to the head variation. The flow is not stationary in the fixed frame inside the rotor, because at any point you can see the rotor blades passing. It is only upstream and downstream of the rotor that the velocities in the fixed frame do not vary over time.

2.1.3.5. *Torque and power*

A second approach[4] to calculate the converted mechanical power uses the angular momentum theorem (section 1.9), which determines the torque applied to the rotor. The mechanical power is deduced from the total torque C_z applied to the rotor by the relationship:

$$\wp_{meca} = C_z.\omega$$

---

3 It is checked in the exercise of section 4.3.

4 This second approach is implemented in the exercises of sections 2.4 (centrifugal pump) and 2.5 (radial flow turbine).

Two formulations of the angular momentum theorem have been introduced in section 1.9.

– When the flow is in a volume of revolution V and it is permanent in the fixed cylindrical frame of reference with axis Oz (this is the case of a stator), the component of the torque along the direction Oz is written as:

$$C_z = -\iint_{S_d} \rho(r\rho u_\theta)(\vec{u}.\vec{n})ds$$

S_d denotes the surfaces surrounding the domain V through which the flow rate is non-zero. Note that $\vec{n} = n_r\vec{e}_r + n_z\vec{e}_z$ is the unit vector perpendicular to these surfaces, oriented outwards from the domain V. This expression means that the torque applied to the solid surface at the boundaries of the domain is equal to the difference between the inflow of angular momentum and the outflow of angular momentum.

– When the volume V is of revolution, but the flow is permanent in the cylindrical reference frame rotating at the angular speed ω around the axis Oz (this is the case of a rotor), the component of the torque according to the direction Oz is written as:

$$C_z = -\iint_{S_d} \rho(r\rho u'_\theta)(\vec{u}'.\vec{n})ds - \iiint_V \rho 2\omega r u'_r dv$$

The expression shows that the Coriolis force in an integral over the volume V, the calculation of which requires knowing the radial component u'_r of the speed.

This triple integral considerably reduces the interest of this approach, but the difficulty can be circumvented in two cases:

– In an axial machine, the radial component of the flow is zero ($u'_r = 0$), the Coriolis term vanishes and the volume integral is therefore zero.

– In a radial flow machine (centrifugal pumps, Francis turbines), the Coriolis term cannot be left out. We present in section 2.1.5 the limiting case of a rotor with an infinite number of blades. The flow can then be considered as permanent both in the fixed frame and in the moving frame. Writing the angular momentum theorem in the fixed frame then makes it possible to determine the torque without having to calculate the integral of the Coriolis term.

2.1.4. *Rotor and stator*

The proper functioning of a roto-dynamic machine generally requires that the rotation of the flow be suppressed at the machine outlet. If this is not the case, the kinetic energy associated with the rotational speed component is not converted into pressure; it is then lost by causing a singular head loss. This is why roto-dynamic machines are often equipped with a stator placed, depending on the case, upstream or downstream of the rotor.

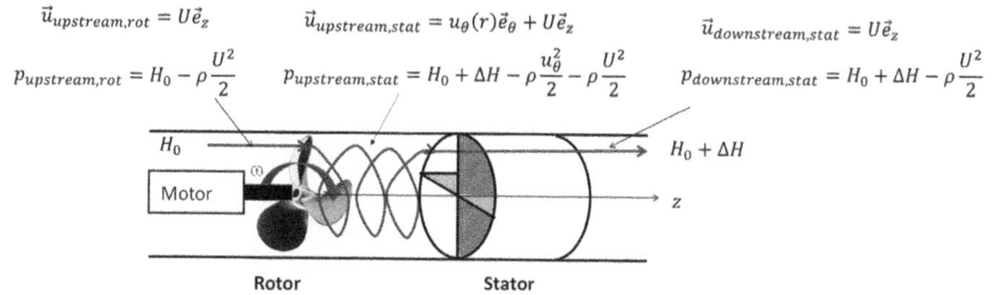

Figure 2.4. *Diagram of an axial pump with a downstream stator. For a color version of this figure, see www.iste.co.uk/mory/mechanical.zip*

The role of the stator is schematized in Figure 2.4 for an axial pump fitted with a stator downstream of the rotor. The axis of rotation of the rotor coincides with the axis Oz of the cylindrical coordinate system in which the speed components are expressed as (u_r, u_θ, u_z). Upstream of the rotor, the fluid does not rotate and the speed is directed along the Oz axis. The rotor generates a speed component $u_\theta(r)$ in the fluid that the stator, placed downstream, aims to eliminate. The stator, consisting of fixed blades, is here a brace of two plates perpendicular to each other and parallel to the axis Oz. It is easy to understand that the flow cannot cross it without considerably modifying its rotation component. When it does so skillfully, that is, without causing head loss, it converts the kinetic energy $\rho u_\theta^2/2$ generated by the rotor into pressure. In the ideal operation schematized in Figure 2.4, the rotor brings the head ΔH to the fluid particles and the pressure varies by the quantity $\Delta H - \rho u_\theta^2/2$ between the inlet and the outlet of the rotor.

In a Francis turbine (see section 2.3), the stator (guide vanes) is placed upstream of the rotor. It generates a rotational component $u_\theta(r)$ by converting pressure into kinetic energy. The rotor recovers this energy and, if well designed, the rotational component is canceled at the rotor output[5].

The volute and the diffuser, at the outlet of a centrifugal pump, are not always equipped with blades to channel the flow, but their function is identical. They convert the rotational kinetic energy at the rotor outlet into pressure, ideally without causing head loss.

2.1.5. *Machine with infinite number of blades*

To consider a stator or a rotor with an infinite number of blades amounts to imagining that there always exists a blade between two neighboring blades. Therefore, the flow is axisymmetric inside the volumes of revolution in which the stator or the rotor is placed and the speed components do not depend on the angle θ. The blades of the stator or of the rotor channel the flow whose speed components in the cylindrical coordinate system are written as $u_r(r,z)$, $u_\theta(r,z)$ and $u_z(r,z)$. The interesting property of the limit where the blades are infinitely close is that, in the case of a rotating rotor, the flow can also be considered as stationary in the fixed frame.

Although viscous effects cannot in principle be neglected when the blades are infinitely close, it is still informative to consider Euler's equations (Navier–Stokes equations for which the viscosity is taken equal to zero) in this limiting case. They are written in the cylindrical reference in the following form:

$$\rho\left\{u_r\frac{\partial u_r}{\partial r}+u_z\frac{\partial u_r}{\partial z}-\frac{u_\theta^2}{r}\right\}=-\frac{\partial p}{\partial r}$$

$$\rho\left\{\frac{u_r}{r}\frac{\partial(ru_\theta)}{\partial r}+\frac{u_z}{r}\frac{\partial(ru_\theta)}{\partial z}\right\}=-\frac{1}{r}\frac{\partial p}{\partial\theta}$$

$$\rho\left\{u_r\frac{\partial u_z}{\partial r}+u_z\frac{\partial u_z}{\partial z}\right\}=-\frac{\partial p}{\partial z}$$

In these equations, p denotes the piezometric pressure. Since the flow is axisymmetric, the derivative $\partial p/\partial\theta$ depends only on r and z. When the angular momentum ru_θ varies along a streamline passing between the blades of a rotor or a stator, the value of $\partial p/\partial\theta$ is non-zero. This pressure gradient translates into a

5 See in section 2.5 the exercise: radial flow turbine model.

pressure difference between the two faces of each blade crossed while traversing a circle of radius r centered on the Oz axis at altitude z. The torque C_z applied to the stator or the rotor results from this non-zero value of $\partial p/\partial\theta$. In the limit where the blades are infinite in number, the force applied to a blade tends toward zero, but the torque C_z applied to the rotor is non-zero because the total surface of the blades tends toward infinity. Euler's equations, in the limit of an infinite number of blades, are therefore another way of understanding the angular momentum theorem[6].

When the rotor has an infinite number of blades, the flow is also permanent in the rotor if it is considered in the fixed frame. It is then advantageous to write the angular momentum theorem while remaining in the fixed frame, since the volume integral of the Coriolis term does not have to be calculated:

$$C_z = -\iint_{S_d} \rho(r\rho u_\theta)(\vec{u}.\vec{n})ds$$

2.1.6. *Similarity of Combe–Rateau*

Roto-dynamic machines verify a similarity law that binds the head ΔH ceded (by a turbine) or gained (by a pump) to the volume flow rate Q passing through the machine. The machine is simply characterized by the diameter D of the rotor and its angular speed of rotation ω. Since the relationship between the parameters stems from a problem of fluid mechanics, the density ρ of the fluid also intervenes in the relationship. On the other hand, we neglected to take into account the viscosity of the fluid, because the flows in the turbomachines being flows with a large Reynolds number, the viscous effects are weak. The relationship therefore links five parameters whose units are presented in Table 2.1.

Settings	ΔH	Q	ω	D	ρ
Units	Pa = kg m^{-1} s^{-2}	m^3 s^{-1}	s^{-1}	m	kg m^{-3}

Table 2.1. *Parameters and units of a roto-dynamic pump*

Involving five parameters with three dimensions (m, kg, s), the relation can, by virtue of the Vashy–Buckingham theorem, be reduced to a relation between two dimensionless numbers whose forms are usually defined as follows:

$$\frac{\Delta H}{\rho\omega^2 D^2} = function\left(\frac{Q}{\omega D^3}\right)$$

6 The exercises in sections 2.4 and 2.5 implement the concepts summarized here.

Blade angles, suction and discharge pipe diameters, etc., are complementary geometric parameters to describe a machine. However, two machines of the same geometry but different sizes will have the same angles and the ratio of any length to rotor diameter will remain the same. This is why it suffices to use a single dimension, here the rotor diameter, to evaluate the size of the machine and to express the relation of similarity.

Figure 2.5 shows the characteristic curve of a roto-dynamic pump. Manufacturers generally indicate a range of flow and head in which the operation of a roto-dynamic pump is recommended (thick continuous line portion of the characteristic curve shown in Figure 2.5). The head ΔH generally decreases as the flow Q increases. Since the power is the product of the flow rate with the head, it then appears that the power (represented as the efficiency curve) is maximum for the nominal flow rate Q_{nom} and the nominal head ΔH_{nom} indicated in the figure.

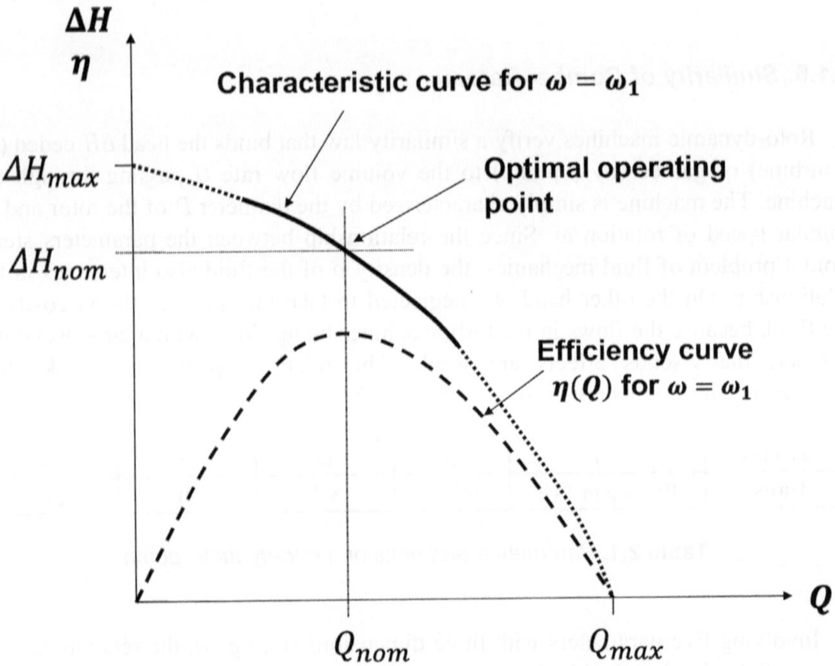

Figure 2.5. *Characteristic curve of a roto-dynamic pump*

It is also common to use the radius of the rotor R rather than its diameter $D = 2R$ to measure the size of a machine.

For the nominal operating conditions of a turbomachine, the head coefficient is defined:

$$\mu = \frac{\Delta H_{nom}}{\rho \omega^2 R^2}$$

and the flow coefficient:

$$\delta = \frac{Q_{nom}}{\omega R^3}$$

Some formulations characterize the rotational speed of the rotor by:

$$n = 60\omega/2\pi$$

expressed in rpm.

2.1.7. *Specific speed and classification of roto-dynamic machines*

For the nominal operating conditions of a roto-dynamic machine, the specific angular speed is used to characterize the technology of a machine:

$$\Omega = \omega \frac{Q_{nom}^{1/2}}{\left(\frac{\Delta H_{nom}}{\rho}\right)^{3/4}}$$

The specific speed also appears in the literature:

$$n_s = n \frac{Q_{nom}^{1/2}}{\left(\frac{\Delta H_{nom}}{\rho g}\right)^{3/4}}$$

with n the rotational speed of the rotor expressed in rpm. The specific speed has the disadvantage of not being a dimensionless number. Thus, the values given in the literature for hydraulic machines vary, especially between documents in French or in English depending on the system of units used. In contrast, the specific angular speed is dimensionless. In the international system of units (kg, m, s), specific angular speed and specific speed are related by the relationship $n_s = 30\,\Omega\, g^{3/4} / \pi$.

Using the expressions for the head coefficient and the flow coefficient introduced in the previous section, we establish that:

$$\Omega = \omega \frac{Q_{nom}^{1/2}}{\left(\frac{\Delta H_{nom}}{\rho}\right)^{3/4}} = \omega \frac{(\delta\omega R^3)^{1/2}}{\left(\frac{\mu\rho\omega^2 R^2}{\rho}\right)^{3/4}} = \frac{\delta^{1/2}}{\mu^{3/4}}$$

The essential property highlighted by this calculation is that the specific angular speed of a machine does not depend on the radius of the rotor R nor on its angular speed of rotation ω. It only depends on the flow coefficient and the head coefficient.

Specific angular speed is used to classify roto-dynamic machines. Schematically, the greater the flow rate required for the same head, the greater the specific speed of the machine.

2.2. Pump dimensioning

2.2.1. *Classification of roto-dynamic pumps*

Pump technologies are classified in Figure 2.6. In the diagrams, the shaded area indicates the rotor and the arrows schematize two streamlines and indicate the direction of the flow. In an axial pump, a streamline remains at the same radial distance r from the axis of rotation of the rotor. In a centrifugal pump or mixed flow pump, the flow enters the machine centrally and the streamlines deviate from the axis of rotation as they pass through the rotor.

Another representation of this classification is shown in Figure 2.7. The area between the two dotted curves indicates the conditions for which pumps can be found at the manufacturers according to their values of the head coefficient and the flow coefficient. The solid curve is the median curve of the relationship $\mu(\delta)$. Choosing a pair of values (δ, μ) located between the two dotted curves means that there is a reasonable chance of finding a pump on the market with these characteristics. Conversely, choosing a pair (δ, μ) outside this domain makes it unlikely that you will find one.

The head coefficient μ varies within a narrow range (between 0.1 and 0.6). It can be noted that the value $\mu = 1$ means that the fluid would have acquired while passing through the rotor the kinetic energy of rotation $\omega^2 R^2/2$ at the periphery of the rotor as well as the associated pressure energy, which has the same value. It would therefore have acquired all the centrifugal energy of the rotor. The maximum value of 0.6 means that we remain below.

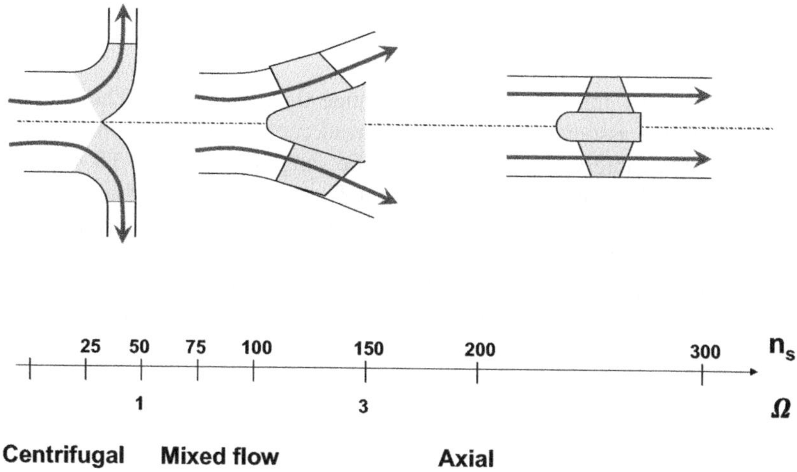

Figure 2.6. *Classification of roto-dynamic pump technologies according to specific angular speed values Ω and specific speed n_s(source: Rey et al. (2012))*

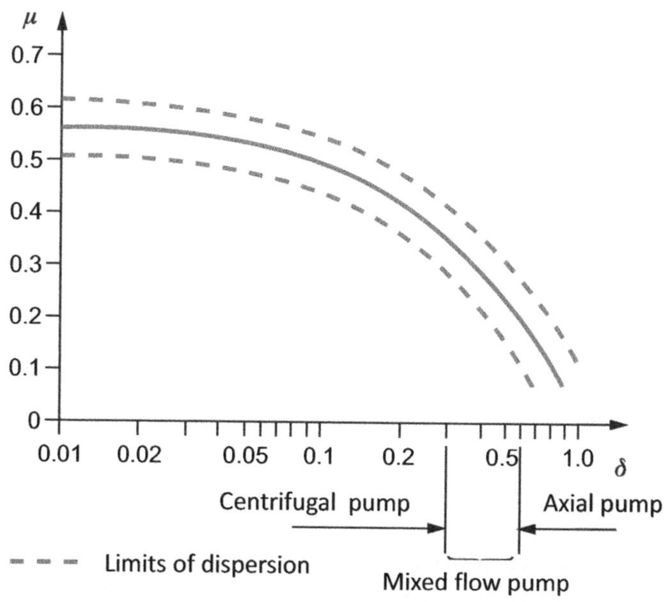

Figure 2.7. *Relationship between the head coefficient and the flow coefficient of roto-dynamic pumps (source: Rey et al. (2012))*

The flow coefficient δ varies over a wide range, which is why the abscissa values are given in a logarithmic scale.

Centrifugal pumps are used in the value range $0.01 < \delta < 0.3$. The bounds of the interval, with the associated values of the head coefficient shown in Figure 2.7, correspond to the interval over the specific angular speed $0.15 < \Omega < 1.1$ (Figure 2.6). For values of the flow coefficient $\delta < 0.01$, corresponding to specific angular speed $\Omega < 0.15$, volumetric pumps are used. These pumps are capable of carrying modest liquid flow rates at high pressure. Their operation consists of inserting the liquid volumes into closed cavities whose volume is reduced during their movement in the pump body. For a liquid, the compressibility of which is very low, the pressure is thus increased by providing very little work. Due to their low use for the energy processes discussed in this work, the subject of volumetric pumps is left aside here[7]. The reader can easily find documentation on the Internet with the keywords volumetric pumps, vane pumps, gear pumps, etc.

A procedure for dimensioning a roto-dynamic pump consists of choosing a pair of values (δ, μ) located between the two dotted curves. By reversing the definitions (section 2.1.6) of the head and flow coefficients, the rotor diameter and the angular speed of rotation are determined according to δ and μ for a nominal head ΔH_{nom} and a Q_{nom} desired nominal flow rate:

$$R = \left(\frac{Q_{nom}}{\delta}\right)^{1/2} \left(\frac{\rho\mu}{\Delta H_{nom}}\right)^{1/4}$$

$$\omega = \left(\frac{\delta}{Q_{nom}}\right)^{1/2} \left(\frac{\Delta H_{nom}}{\rho\mu}\right)^{3/4}$$

Decreasing the value of the flow coefficient δ requires increasing (in a limited proportion) the head coefficient μ. The previous relations indicate that the radius of the rotor will increase and its rotational speed will decrease. Figure 2.7 also indicates the technology of the chosen pump (centrifugal, helico-centrifugal or axial), depending on the value of the flow coefficient δ.

2.2.2. Centrifugal pumps

The scheme of a centrifugal pump is shown in Figure 2.8. Let us recall the direction of the flow which enters through the tube centered on the axis of rotation and exits through the diffuser. The rotor (the darker element) centrifuges the fluid.

---

7 The osmotic energy process studied in section 6.2 is the only one in this work that requires the use of volumetric pumps.

At the exit of the rotor, the relative speed (in the rotating frame of the rotor) is tangent to the trailing edge of the blades in application of Joukovski's condition. The relative speed is represented with a modulus lower than that of the peripheral speed of rotation $\omega R\vec{e}_\theta$ of the rotor because, as we will see in the exercise proposed in section 2.4, the relative speed decreases as we move away from the axis of rotation.

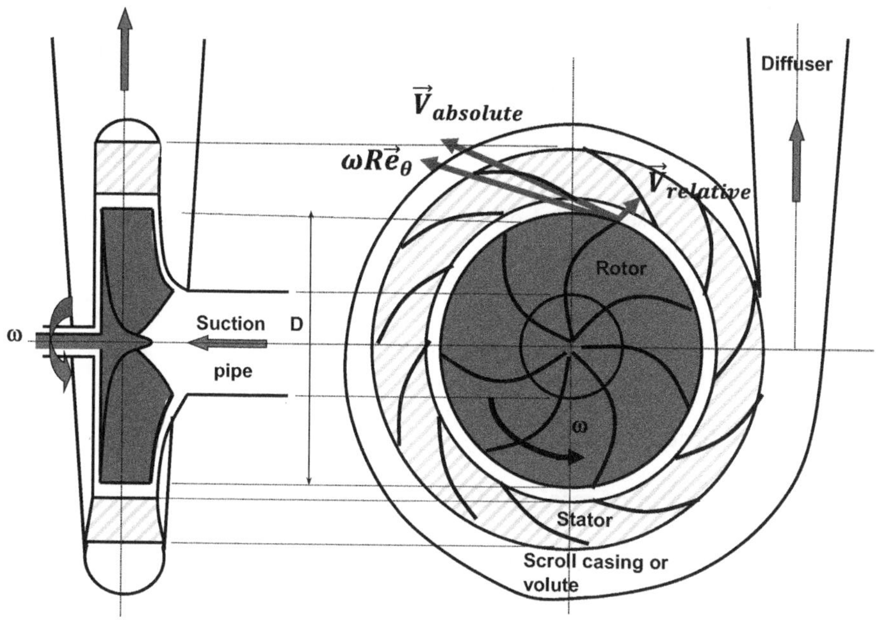

Figure 2.8. *Diagram of a centrifugal pump*

In the fixed frame, the fluid is accelerated as it passes through the rotor. The pressure is also increased to balance the centrifugal force applied to the fluid particle. These reasons explain the name centrifugal pump.

Dimensioning of the volute and diffuser should be done carefully. If the inlet section and the outlet section are identical, the delivery speeds at the inlet and at the outlet of the pump are identical. The slowing down of the fluid, in the volute and the diffuser must be carried out gradually to avoid a head loss.

Figure 2.8 outlines the successive conversion steps carried out in a centrifugal pump. The electric motor that drives the rotor first transmits the consumed electrical energy to the fluid that passes through the rotor, increasing its pressure energy and its kinetic energy. Then the volute and the diffuser slow down the fluid, converting some of the kinetic energy of the fluid into pressure energy.

Starting a centrifugal pump can be tricky. Filled with air, the pump cannot produce sufficient pressure decrease to pump water from a pool if it is placed too high above it. To start a centrifugal pump, the body of the pump must first be filled with water and the aim is to maintain a volume of water within it. Furthermore, when the body of the pump is filled with air while the pump is intended to pump water, the torque applied by the flow of air on the rotor cannot equal the torque applied by the electric motor. This situation can burn out the electric motor if it continues.

In the exercise of section 2.4 of this chapter, we carry out the dimensioning of a centrifugal pump. This exercise, with an academic approach, implements the concepts introduced in section 2.1. In addition, in Chapter 5 (on hydroelectricity), the exercise in section 5.3 involves the dimensioning of a pump with an engineer's approach, which uses Figure 2.7.

2.3. Turbine dimensioning

2.3.1. *Classification of roto-dynamic turbines*

Turbine technologies are classified in Figure 2.9 according to specific angular speed. The external appearances of a Francis turbine and a centrifugal pump are similar. They are distinguished first by the direction of flow. The fluid enters the Francis turbine through the volute (arrow in Figure 2.9) and exits through the central orifice, unlike what happens in a centrifugal pump (Figure 2.8). Figure 2.10 shows the interior of a Francis turbine. The blades of the stator, otherwise called guide vanes, placed between the volute and the rotor, generate a rotational component u_θ of the flow, converting the pressure energy of the fluid into kinetic energy. In ideal operation, the rotor suppresses the rotational speed component u_θ generated by the guide vanes. It recovers part of the kinetic energy and the pressure energy of the fluid which the alternator, driven by the rotor, converts into electrical energy. The angle of the guide vanes must take care not to generate an excessive component of rotation, so as not to excessively reduce the pressure and avoid the appearance of cavitation. A Francis turbine is a radial flow machine. The rotor geometry of a Francis turbine is complex and differs significantly from that of a centrifugal pump. In the exercise of section 2.5, we study the dimensioning of a radial flow turbine in a simplified geometry compared to that of a Francis turbine. This exercise applies the concepts of section 2.1. It also makes it possible to understand the role of the stator and to consider blade shapes for the rotor, depending on the head and flow conditions. It also studies the risk of cavitation.

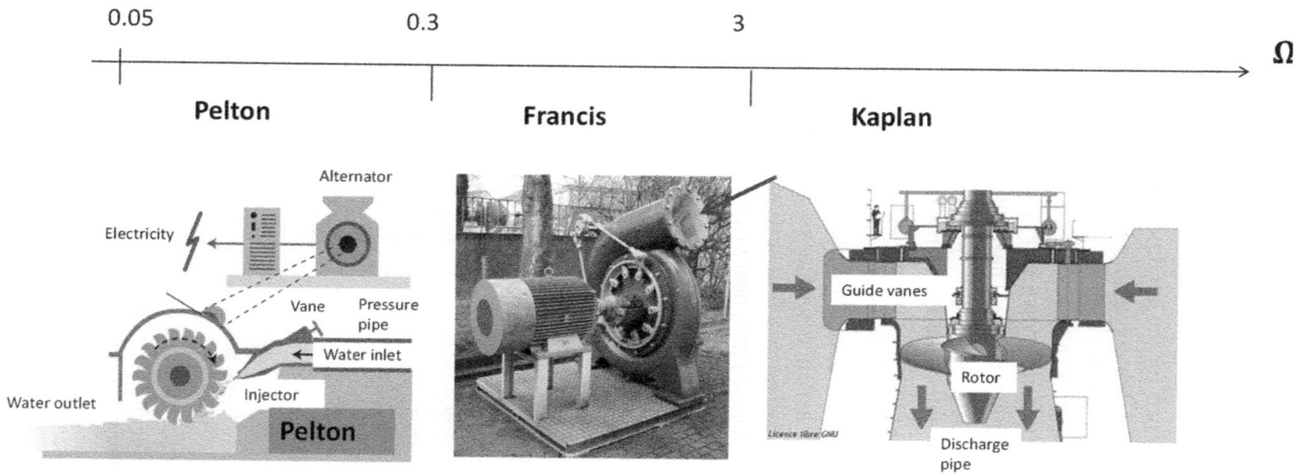

Figure 2.9. *Classification of roto-dynamic turbine technologies according to the specific angular velocity Ω (source: Megnint et al. (2008); https://lenergie-solaire.net/energies-renouvelables/energie-hydraulique/turbines-hydrauliques/turbine-francis; http://moulindewanlin.be/la-centrale/les-principaux-types-de-turbine/; https://www.pinterest.fr/pin/180495897546526042/)*

The Kaplan turbine is similar to an axial pump. These two machines differ essentially by the orientation of their blades. In a turbine, the fluid pushes the blades in the direction of rotation of the rotor, whereas in a pump the flow opposes the rotation of the rotor. Although the Kaplan turbine has a radial inlet in Figure 2.9, the flow discharging into the rotor is axial. In the exercise proposed in section 4.3, a Kaplan turbine is dimensioned for the conditions of the La Rance tidal power plant. The dimensioning principles are common with the dimensioning of the rotor of a wind turbine (exercise in section 3.3).

The third category of turbine, shown in Figure 2.9, is the Pelton turbine, which has no equivalent in pumping mode. The block diagram included in this figure summarizes its main features. The Pelton turbine operates at atmospheric pressure. The first element of the turbine is the injector which transforms pressure energy into kinetic energy. The Pelton turbine is often used in high head conditions and the jet speed exiting the injectors is very high. The jet from the injectors impacts buckets placed on the periphery of a wheel which rotates by activating an alternator. The dimensioning principle of a Pelton turbine is presented in the next section. This model is used in the exercise proposed in section 5.3 to size the Pelton turbines of the Grand'Maison hydroelectric plant, near Grenoble.

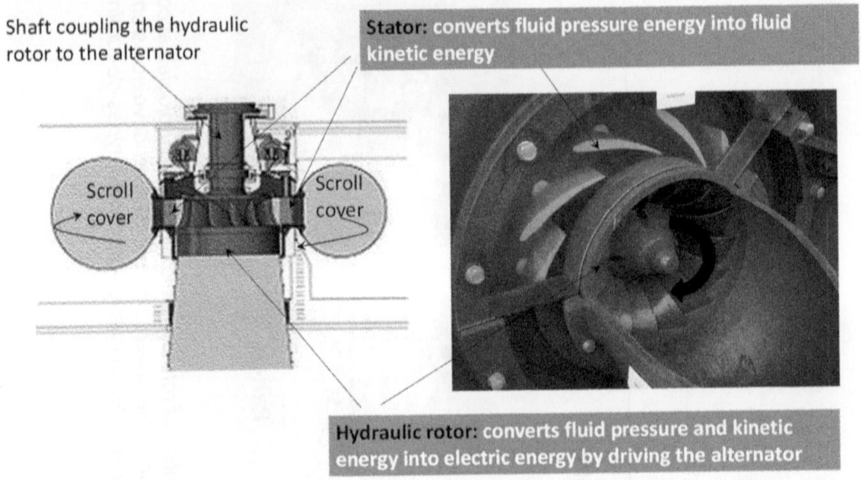

Figure 2.10. *Interior elements of a Francis turbine: volute, rotor (dark) and stator or distributor (light)*[8]. *For a color version of this figure, see www.iste.co.uk/mory/mechanical.zip*

8 Available at: https://en.wikipedia.org/wiki/Turbine_Francis, Wikimedia Commons (CC BY-SA 3.0).

Turbine	Pelton	Francis	Kaplan
Drop height (m)	200–2,000	30–300	2–25
Flow (m^3/s)	4–15	10–700	70–800

Table 2.2. *Use of Pelton, Francis and Kaplan turbines depending on head and flow conditions*

In hydroelectricity (Chapter 5), the fields of implementation of the various turbines are generally chosen according to the values presented in Table 2.2. These are the conditions for which they are most often used. However, in the context of micro-hydroelectric installations, the use of a Pelton turbine for much lower head heights can theoretically be envisaged. The flow rate is reduced, and the calculated specific speed may also be based on a Pelton-type technology. For this reason, some popularizers rightly describe the watermill as the ancestor of the Pelton turbine, but it has the disadvantage that the wheel diameter is large, and the rotational speed is slow for these operating conditions.

Finally, we recommend watching a few video clips that illustrate the technologies of hydraulic turbines:

– on the use of turbines[9];

– on Pelton turbines[10];

– on Francis turbines[11];

– on Kaplan turbines[12].

2.3.2. *Dimensioning of Pelton turbines*

Figure 2.11 shows a more detailed diagram of a Pelton turbine than Figure 2.9. In particular, it specifies its location within a hydroelectric facility. The difference in altitude H between the surface of the upstream reservoir and the turbine must be imagined to be very large in relation to the size of the turbine. The diagram highlights the role of the penstock that brings water from the upstream reservoir to the turbine inlet: it converts the potential energy of the fluid into pressure energy at the turbine inlet. The second stage of the conversion is carried out by the injectors (the turbine here is equipped with six injectors): they convert the pressure energy

9 Available at: https://www.youtube.com/watch?v=k0BLOKEZ3KU.

10 Available at: https://www.youtube.com/watch?v=r5GPkATPeXk.

11 Available at: https://www.youtube.com/watch?v=3BCiFeykRzo.

12 Available at: https://www.youtube.com/watch?v=0p03UTgpnDU.

into kinetic energy of the jets which emerge into the open air (and therefore at atmospheric pressure). Finally, the third stage of the conversion is carried out by the rotor which rotates under the action of the jets impacting the buckets. It drives the alternator which converts the kinetic energy of the jets into electrical energy.

Figure 2.11 also shows the photograph of a Pelton turbine wheel as well as the detail image of the buckets. They have the shape of a double-shelled walnut. The jet that impacts the bucket splits in two on the central edge and the fluid is ejected by turning around on the side faces of the troughs. This is well illustrated in the animations of the clip mentioned in the previous section.

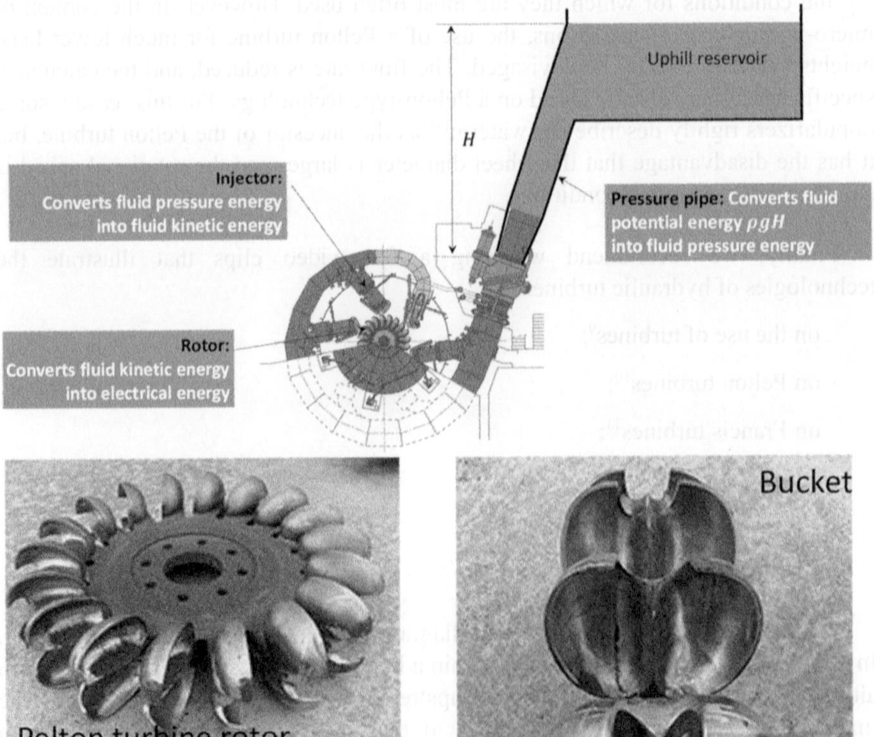

Figure 2.11. *Diagram of a Pelton turbine and its layout. Photographs of a wheel and details of buckets*[13]

13 Available at: https://en.wikipedia.org/wiki/Turbine_Pelton.

The energy conversion model by a Pelton turbine assumes that the buckets are located in large numbers on the periphery of the wheel, as shown in the photograph in Figure 2.11. For this reason, the buckets are notched in their upper portion so as to allow the jet issuing from an injector to properly impact the buckets successively.

The first dimensioning step concerns the injectors. The head H_e at the turbine inlet is related to the head h by the relationship:

$$H_e = P_{atm} + \rho g h - \Delta H = P_{atm} + \rho g h_{eff}$$

ΔH designates the head loss on the portion of the streamline between the surface of the upper reservoir and the inlet of the turbine, usually the linear head loss in the penstock (which one seeks to keep small). The operators introduce the effective height h_{eff} equal to the height of fall minus this head loss. Between the inlet of the turbine and the outlet of the injectors, the fluid is accelerated. The speed of the jet leaving the injectors is determined by applying Bernoulli's theorem between these two points:

$$u_1 = \sqrt{2gh_{eff}}$$

We recognize in this formula the Torricelli formula (see section 1.2).

The flow section at the outlet of the injectors is generally a circular disk. Its diameter d_{inj} can be changed to adjust the flow rate. For a Pelton turbine with a number n_{inj} of injectors, the flow rate passing through the turbine is:

$$Q = n_{inj} u_1 \frac{\pi d_{inj}^2}{4}$$

Conversely, the diameter of the injectors is deducted from the flow rate value:

$$d_{inj} = \sqrt{\frac{4Q}{\pi u_1 n_{inj}}}$$

Within the limit of a rotor with an infinite number of blades (buckets), the flow can be considered as stationary in the fixed frame. It is within this limit that the wheel of the Pelton turbine is dimensioned. Figure 2.12 shows the configuration. It is easily understood that the diameter of the wheel must be large enough with respect to the dimensions of a bucket and with respect to the diameter of the jet coming from the injectors to be able to plant a large number of buckets on the wheel. The diameter D_{rot} of the rotor is quantified by the form factor which, for Pelton turbines, is in the interval:

$$9 < \frac{D_{rot}}{d_{inj}} < 30$$

With the three previous formulas, we determine successively the jet speed, the diameter of the jets and the diameter of the rotor of a Pelton turbine for an effective height of fall h_{eff} and a known flow rate Q.

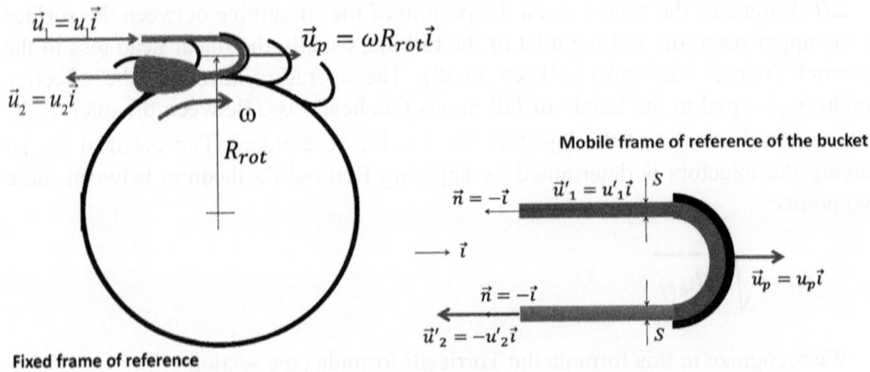

Figure 2.12. *Diagram of application of the momentum theorem for a Pelton turbine wheel*

The final dimensioning step determines the optimal angular speed of rotation of the wheel. This is done by applying the momentum theorem to the wheel of the Pelton turbine, based on the diagram in Figure 2.12. Since there are an infinite number of buckets, it can be done in the fixed frame because the flow is considered permanent there. The vector $\vec{i}$ is the unit vector in the direction of the jet exiting the injector. Note that u_2 designates the component of the speed (in the fixed frame) at the outlet of the buckets according to the direction $\vec{i}$ and the speed of the bucket at the point of impact of the jet is $\vec{u}_p = \omega R_{rot}\vec{i}$. At this point, only the rotational angular speed ω and speed u_2 are undetermined.

Considering that at any instant an injector jet impacts a single bucket, the momentum theorem (in the fixed frame) is applied to the volume of water indicated in blue in Figure 2.12. We deduce the force applied by the flow on the trough:

$$\vec{F}_{bucket} = \rho Q_{bucket} u_1 \vec{i} - \rho Q_{bucket} u_2 \vec{i}$$

The first term to the right of the equality is the momentum flux on the incoming surface and the second on the outgoing surface. In this expression, it is recommended to show the flow rate Q_{bucket} impacting the bucket. It is equal to the product of the speed $u_1$with the surface of the jet issuing from the injector, but the flow rate cannot be linked to the speed u_2 because the flowing surface at the outlet is not known. In this calculation, the pressure does not intervene, since the Pelton turbine operates at atmospheric pressure.

The mechanical power recovered by the Pelton turbine via the considered bucket is:

$$\wp_{bucket} = \vec{F}_{bucket}\vec{u}_p = \{\rho Q_{bucket}u_1\vec{\imath} - \rho Q_{bucket}u_2\vec{\imath}\}\omega R_{rot}\vec{\imath}$$

By summing for the number of injectors, $Q = n_{inj}Q_{bucket}$ and the total power recovered by the turbine is:

$$\wp = n_{inj}\wp_{bucket} = \rho Q\{u_1 - u_2\}\omega R_{rot}$$

This expression always includes two unknowns, the angular speed of rotation ω and the speed u_2.

Under the assumption of ideal operation of the turbine, that is, assuming that any energy released by the fluid is recovered by the turbine and that no additional head loss interferes with this balance, it is possible to determine the value of u_2.

By placing oneself in the frame moving with the bucket, the relative velocities in this frame at entry and exit are by the composition of the absolute and relative velocities:

$$u'_1\vec{\imath} = u_1\vec{\imath} - \omega R_{rot}\vec{\imath}$$

$$-u'_2\vec{\imath} = u_2\vec{\imath} - \omega R_{rot}\vec{\imath}$$

The negative sign in the second expression is introduced to keep a value of u'_2 positive. If the energy conversion goes well, that is to say without parasitic head loss, the head must be conserved if it is calculated by evaluating the kinetic energy of the flow in the relative reference. The pressure plays no role since the process is at atmospheric pressure. The kinetic energy of the flow in the relative frame is conserved and therefore $u'_1 = u'_2$. We therefore deduce from the previous expressions that:

$$u_1 - \omega R_{rot} = \omega R_{rot} - u_2$$

Using this relation, the speed u_2 is then eliminated from the expression for the power:

$$\wp = \rho Q 2\{u_1 - \omega R_{rot}\}\omega R_{rot}$$

The choice of the angular speed of rotation ω makes it possible to maximize the power recovered. It is clear that the power recovered is zero if the wheel is not turning. At the other end of the possible conditions, if the wheel is spinning at the same speed as the jet, there is also no recuperated speed (it is well known that a rugby player has little impact on his opponent if he runs at the same speed). The power recovered is maximum if:

$$\frac{d\wp}{d\omega} = 0$$

that is, if the speed of the bucket is half the speed of the jet:

$$\omega R_{rot} = \frac{u_1}{2}$$

The power recovered by the Pelton turbine is then given as:

$$\wp_{Pelton} = Q\rho \frac{u_1^2}{2}$$

Finally, it is verified that $u_2 = 0$, which means that all of the kinetic energy of the jets has been recovered by the turbine and that its efficiency (power relative to the kinetic energy of the jet multiplied by the flow passing through the turbine) is equal to 1.

It is usual to evaluate the speed of rotation of the wheel in number of revolutions per minute, that is:

$$n = 60\frac{\omega}{2\pi} = \frac{15}{\pi}\frac{u_1}{R_{rot}}$$

A Pelton turbine dimensioning is carried out in the exercise of section 5.3 for the Grand'Maison hydroelectric plant.

2.4. Exercise: centrifugal pump model

The impeller of a centrifugal pump, with six helical-shaped blades, is shown in Figure 2.13. The fluid is introduced through the central orifice with axis Oz. The rotor spins at angular speed ω, counterclockwise. The blades channel the fluid by centrifuging it between two planes perpendicular to the Oz axis, separated by a distance h that corresponds to the height of the blades.

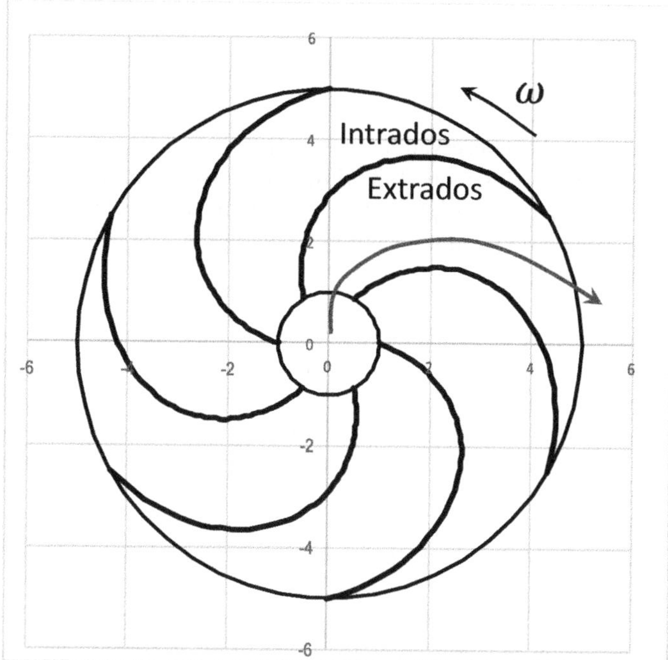

Figure 2.13. *Diagram of a centrifugal pump rotor. The arrowed line indicates the trajectory of a fluid particle*

The shape of the blades, described in the cylindrical coordinate system with axis Oz by its polar coordinates (r, θ), is given by the vector:

$$\overrightarrow{OM} = r(\theta)\vec{e}_r$$

The tangent vector to the blades is given as:

$$\frac{d\overrightarrow{OM}}{d\theta} = \frac{dr}{d\theta}\vec{e}_r + r\vec{e}_\theta$$

In the exercise, we will choose to describe the blade by the function $\theta(r)$ rather than using the function $r(\theta)$. The leading edge of the reference blade is located at $(\theta = 0, r = R_i)$ and the trailing edge is at $(\theta = \theta_{max}, r = R_e)$. The angle θ_{max} will be determined during dimensioning.

This centrifugal pump model considers a rotor with an infinite number of blades. The vector $\overrightarrow{OM} = r(\theta)\vec{e}_r$ describing a blade then also traces the streamline followed by a fluid particle. Since the speed vector of the flow in the rotating frame is tangent to the streamline:

$$u'_r(r) = \alpha \frac{dr}{d\theta} \quad \text{and} \quad u'_\theta(r) = \alpha r$$

1) When writing the conservation of the flow rate (through a surface to be specified), express α as a function of the flow rate Q passing through the rotor. Then express the speed components $u'_r(r)$ and $u'_\theta(r)$ in the rotating frame, as well as the components of speed $u_r(r)$ and $u_\theta(r)$ in the fixed frame. We will take into account that the blade is described by the function $\theta(r)$.

2) The rotation of the blades imparts a rotational movement to the fluid particles channeled between the rotor blades. The fluid entering the rotor does not rotate. Write the necessary condition on the shape of the blade $\theta(r)$ in $r = R_i$. Determine the function $\theta(r)$ in the interval $R_i < r < R_e$ for which the component $u_\theta(r)$ of the flow in the fixed frame is zero throughout the rotor. Give a kinematic interpretation of this flow.

3) Use the angular momentum theorem to determine the component of torque C_z applied by the flow on the rotor. As the blades are infinite in number, the flow in the rotor is permanent in the fixed frame as in the rotating frame. We will verify that the angular momentum theorem gives the same result depending on whether it is applied in the fixed frame or in the rotating frame and we will explain the Coriolis term when the angular momentum theorem is applied in the rotating frame.

4) Calculate the electric power consumed by the pump and the head brought to the fluid by the rotor. Indicate the conditions to be verified for the machine to operate in pumping mode.

5) The nominal operating conditions are obtained for the flow rate Q corresponding to the maximum power. Determine the nominal flow rate Q_{nom} and the nominal head ΔH_{nom} supplied to the fluid. Comment on the results.

6) It is assumed that the radius R_c of the fluid inlet pipe (oriented along the axis of the rotor) is equal to the inside radius R_i of the rotor. Note R_e is its outer radius. The geometry of the rotor (Figure 2.13) is such that $R_e = 5R_i$ and $R_i = R_c$. We want to size a pump with nominal head $\Delta H_{nom} = 3$ bar and nominal flow rate $Q_{nom} = 30$ L/s. The rotor rotates at speed $n = 1{,}200$ rpm. The fluid is water. Determine the diameters R_e and R_i. Calculate the height h of the blades for which the speed of the fluid at the inlet of the rotor ($r = R_i$) is equal to the speed in the

suction pipe of the pump. Calculate the flow rate coefficient $\delta = Q_{nom}/\omega R_e^3$ of the pump and comment on the numerical values.

7) Calculate the shape $\theta(r)$ of the reference blade allowing the torque $C_z(r)$ applied to the rotor element between the radius cylinder R_i and the radius cylinder r to increase linearly with r.

8) Euler's equations in a cylindrical coordinate system are written, with p designating the piezometric pressure:

$$\rho\left\{\frac{\partial u_r}{\partial t} + u_r\frac{\partial u_r}{\partial r} + \frac{u_\theta}{r}\frac{\partial u_r}{\partial \theta} + u_z\frac{\partial u_r}{\partial z} - \frac{u_\theta^2}{r}\right\} = -\frac{\partial p}{\partial r}$$

$$\rho\left\{\frac{\partial u_\theta}{\partial t} + \frac{u_r}{r}\frac{\partial (r u_\theta)}{\partial r} + \frac{u_\theta}{r}\frac{\partial u_\theta}{\partial \theta} + u_z\frac{\partial u_\theta}{\partial z}\right\} = -\frac{1}{r}\frac{\partial p}{\partial \theta}$$

$$\rho\left\{\frac{\partial u_z}{\partial t} + u_r\frac{\partial u_z}{\partial r} + \frac{u_\theta}{r}\frac{\partial u_z}{\partial \theta} + u_z\frac{\partial u_z}{\partial z}\right\} = -\frac{\partial p}{\partial z}$$

After recalling the expressions for the speed components in the rotor, write the previous equations keeping only the non-zero terms and comment on the result.

Calculate the pressure difference $\{p(R_e) - p(R_i)\}$ by integrating the pressure variation along a streamline, that is, by integrating:

$$\frac{\partial p}{\partial r} + \frac{\partial p}{\partial \theta}\frac{d\theta}{dr}$$

Why is it not possible to integrate Euler's equation along the radial direction alone?

Express the head variation depending on whether the speed components used are those in the fixed frame or those in the rotating frame. Comment on the results obtained with a rotor rotating at the angular speed ω depending on the flow rate being zero ($Q = 0$) or that the flow rate is equal to the nominal flow rate ($Q = Q_{nom}$).

Carry out the numerical applications for the operating conditions given in question 6. Summarize in a table the numerical values of the speed components $u_r(R_i)$, $u_\theta(R_i)$, $u_r(R_e)$ and $u_\theta(R_e)$ in the fixed frame as well as their values $u'_r(R_i)$, $u'_\theta(R_i)$, $u'_r(R_e)$ and $u'_\theta(R_e)$ in the rotating frame. Calculate the variations of the kinetic energy of the radial velocity component and the kinetic energy of the azimuthal velocity component between the entry and the exit of the rotor.

9) To what extent can the results of this exercise be extrapolated for a rotor with a finite number of blades?

10) Figure 2.14 shows the complete layout diagram of a centrifugal pump equipped with the rotor dimensioned in the previous questions. This pump draws water from a pond. It is assumed that the head at the pump outlet is such that the pump operates with its nominal flow Q_{nom} and head conditions ΔH_{nom}. The suction pipe is assumed to be short enough to be able to neglect the linear head loss in it.

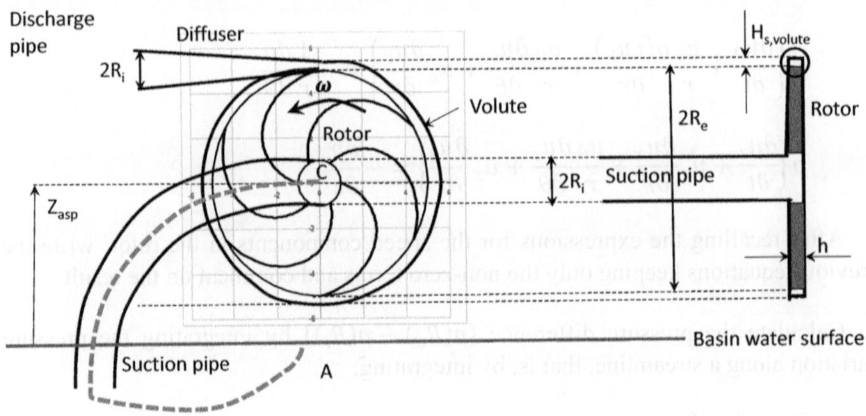

Figure 2.14. *Diagram of the layout of the centrifugal pump (suction pipe, rotor, volute then diffuser) to pump water from a pool*

Calculate the pressure at the inlet of the pump placed at a height Z_{asp} above the free surface of the pool. Cavitation occurs when the pressure is lower than the vapor pressure of water (P_v = 2,300 Pa for liquid water at a temperature of 20°C). Liquid water is then vaporized and the flow becomes two-phase. This situation should be avoided. In terms of energy, it reduces the performance of the pump. Moreover, cavitation is liable to damage the solid walls of the blades during the collapsing of air bubbles on a solid surface, which generates violent impacts on this surface. Determine the maximum value of the suction height Z_{asp}.

11) Finally, we carry out the dimensioning of the elements downstream of the rotor of the pump shown schematically in Figure 2.14. The volute collects the flow from the rotor to send it to the outlet pipe of the pump. We saw in question 8 that the kinetic energy of the flow is greater at the outlet of the rotor than its value at the

inlet. It is therefore necessary to slow down the fluid to get it out of the pump via a delivery pipe of the same section as that of the suction pipe. This is what the diffuser is for. The widening of the pipe must be gradual enough to convert the kinetic energy into pressure without causing pressure loss.

The volute has a rectangular section of height h (in the Oz direction) equal to the height of the rotor blades. Its thickness $H(\theta)$ in the radial direction increases from the start of the volute $H(\theta = 0) = 0$ until its value at the volute outlet $H(\theta = 2\pi) = H_{s,volute}$. The widening of $H(\theta)$ is carried out in such a way as to keep the azimuthal $u_\theta(R_e)$ component in the volute constant. Deduce the function $H(\theta)$ and its value $H(\theta = 2\pi) = H_{s,volute}$ at the volute output.

2.4.1. *Solution*

1) The conservation of the flow rate through any circular cylinder of radius r and height h (height of the blade along the Oz axis), $Q = 2\pi rhu'_r$ makes it possible to connect the components of speed in the rotating frame to the shape of the blades:

$$u'_r(r) = \frac{Q}{2\pi h}\frac{1}{r} = \alpha\frac{dr}{d\theta}$$

$$u'_\theta(r) = \alpha r = \frac{Q}{2\pi h}\frac{1}{\frac{dr}{d\theta}} = \frac{Q}{2\pi h}\frac{d\theta}{dr}$$

When the rotor works as a pump, the radial component $u'_r(r)$ is positive.

The speed components in the fixed frame are then deduced by adding the rotation component of the rotor:

$$u_r(r) = \frac{Q}{2\pi h}\frac{1}{r} \quad \text{and} \quad u_\theta(r) = \frac{Q}{2\pi h}\frac{d\theta}{dr} + \omega r$$

The speed components are axisymmetric (they do not depend on θ) for a rotor with an infinite number of blades.

2) Since the flow does not turn in the fixed frame at the inlet of the rotor, it comes:

$$u_\theta(R_i) = \frac{Q}{2\pi h}\frac{d\theta(R_i)}{dr} + \omega R_i = 0$$

The derivative at the leading edge of the blade is therefore:

$$\frac{d\theta(R_i)}{dr} = -\frac{2\pi h\omega R_i}{Q}$$

If the rotational component is zero throughout the rotor, then for any radial position r:

$$u_\theta(r) = \frac{Q}{2\pi h}\frac{d\theta}{dr} + \omega r = 0$$

The integration makes it possible to determine the shape of the reference blade (such as $\theta = 0$ for $r = R_i$), which corresponds to this situation:

$$\theta(r) = -\frac{\pi h\omega}{Q}(r^2 - R_i^2)$$

For this shape of blade, the fluid particles slide on the blade without undergoing any force which deflects their trajectory. The azimuthal motion in the rotating frame due to the blade, associated with the radial flow, adjusts at any position r to compensate for the rotational component of the rotor ωr. The rotor does not transfer any head to the fluid particles passing through it. The flow is reduced to the radial flow that one would have in the absence of the rotating rotor.

3) In the fixed frame, the torque applied by the flow on the rotor is calculated by the angular momentum theorem (see sections 1.9 and 2.1.5). It is written using the speed components explained in question 1:

$$C_z = Q\{R_i\rho u_\theta(R_i) - R_e\rho u_\theta(R_e)\}$$
$$= \rho Q\left\{\frac{Q}{2\pi h}\frac{d\theta(R_i)}{dr}R_i + \omega R_i^2 - \frac{Q}{2\pi h}\frac{d\theta(R_e)}{dr}R_e - \omega R_e^2\right\}$$

In the moving frame, the angular momentum theorem is written as:

$$C_z = Q\{R_i\rho u'_\theta(R_i) - R_e\rho u'_\theta(R_e)\} - \iiint_V \rho 2\omega r u'_r dv$$
$$= Q\{R_i\rho u'_\theta(R_i) - R_e\rho u'_\theta(R_e)\} - \rho 2\omega\frac{Q}{2\pi h}\iiint_V dv$$
$$= \rho Q\left\{\frac{Q}{2\pi h}\frac{d\theta(R_i)}{dr}R_i - \frac{Q}{2\pi h}\frac{d\theta(R_e)}{dr}R_e\right\}$$
$$-\rho 2\omega\frac{Q}{2\pi h}\{h\pi(R_e^2 - R_i^2)\}$$

We obtain the same result. In the moving frame, we can integrate the Coriolis force on the volume because the blades are infinite in number and the speed $u'_r(r)$ is known.

4) Since the rotational component in the fixed frame is zero at the rotor inlet (question 1), the expression of the torque applied by the flow on the rotor is simplified to:

$$C_z = -\rho Q \left\{ \frac{Q}{2\pi h} \frac{d\theta(R_e)}{dr} R_e + \omega R_e^2 \right\}$$

By the principle of action-reaction, the opposite torque must be applied by the electric motor to make the rotor turn. The power supplied to the fluid is therefore:

$$\wp = -C_z \omega = \Delta H_{pump} Q = \rho Q \omega \left\{ \frac{Q}{2\pi h} \frac{d\theta(R_e)}{dr} R_e + \omega R_e^2 \right\}$$

The head supplied to the fluid by the pump is therefore:

$$\Delta H_{pump} = \rho \omega \left\{ \frac{Q}{2\pi h} \frac{d\theta(R_e)}{dr} R_e + \omega R_e^2 \right\}$$

The machine operates in pump mode if $\Delta H_{pump} > 0$. It is a turbine if the inequality is in the opposite direction.

5) The maximum power is obtained for the nominal flow such that $d\wp/dQ = 0$, which leads to:

$$Q_{nom} = -\frac{\pi h \omega R_e}{\frac{d\theta(R_e)}{dr}}$$

$$\Delta H_{nom} = \rho \frac{\omega^2 R_e^2}{2}$$

The head coefficient is therefore:

$$\mu = \frac{\Delta H_{nom}}{\rho \omega^2 R_e^2} = \frac{1}{2}$$

This value is in agreement with the values obtained for centrifugal pumps (Figure 2.7).

6) Since the rotational speed is known (it is fixed by the electric motor which drives the rotor), the outer diameter R_e is chosen to ensure that the pump delivers the nominal head:

$$R_e = \sqrt{\frac{2\Delta H_{nom}}{\rho\omega^2}}$$

The inside radius of the rotor R_i and the radius R_c of the suction pipe are deduced from the geometric relations $R_e = 5R_i$ and $R_i = R_c$ fixed by the statement.

The equality between the flow rate in the suction pipe (speed U_{asp}) and the flow passing through the cylinder of radius R_i and height h is written as:

$$Q_{nom} = 2\pi R_i h\, u'_r(R_i) = U_{asp}\pi R_i^2$$

The rotor dimensions are chosen so that $u'_r(R_i) = U_{asp}$. Therefore, $h = R_i/2$. This choice avoids a modification of the speed of the fluid at the inlet of the pump, which could cause a loss of head or a reduction in pressure liable to cause cavitation.

The pump flow coefficient is calculated by:

$$\delta = \frac{Q_{nom}}{\omega R_e^3}$$

The numerical results of the dimensioning are given in Table 2.3.

The outer radius R_e of the rotor is reasonable. The thickness h is low. The flow coefficient δ is in the range of centrifugal pumps (Figure 2.7).

ΔH_{nom} (Pa)	Q_{nom} (m^3/s)	n (rpm)	ω (rad/s)	R_e (m)	δ (–)	R_i (m)	h (m)
$3.0\ 10^5$	0.03	1,200	125.7	0.195	0.032	0.039	0.019

Table 2.3. *Geometric dimensions and flow coefficient of the pump*

7) The blade angle at the trailing edge is deduced from the expression of the nominal flow rate obtained in question 5:

$$\frac{d\theta(R_e)}{dr} = -\frac{\pi h\omega R_e}{Q_{nom}}$$

The torque applied to the rotor is then, using the expression of question 4:

$$C_z(R_e) = -\rho Q_{nom} \frac{\omega R_e^2}{2}$$

The shape of the blades is chosen so that the torque $C_z(r)$ applied to the rotor element between the cylinder of radius R_i and the cylinder of radius r increases linearly with r. This choice makes it possible to evenly distribute the forces on the rotor. This hypothesis translates into:

$$C_z(r) = -\rho Q_{nom} \frac{\omega R_e^2}{2} \frac{r - R_i}{R_e - R_i} = -\rho Q_{nom} \left\{ \frac{Q_{nom}}{2\pi h} \frac{d\theta}{dr} r + \omega\, r^2 \right\}$$

The differential equation governing the shape of the blade is therefore written as:

$$\frac{d\theta}{dr} = -\frac{\omega 2\pi h}{Q_{nom}} r + \frac{\omega R_e^2}{Q_{nom}} \frac{\pi h}{R_e - R_i} \left\{ 1 - \frac{R_i}{r} \right\}$$

Its integration is easy:

$$\theta(r) = -\frac{\omega R_e^3}{Q_{nom}} \frac{\pi h R_i^2}{R_e^3} \left(\frac{r^2}{R_i^2} - 1 \right) + \frac{\omega R_e^3}{Q_{nom}} \frac{\pi h}{(R_e - R_i)} \frac{R_i}{R_e} \left\{ \frac{r}{R_i} - 1 - Ln\left(\frac{r}{R_i} \right) \right\}$$

The angle θ_{max} at the trailing edge is:

$$\theta_{max} = -\frac{\omega R_e^3}{Q_{nom}} \frac{\pi h R_i^2}{R_e^3} \left(\frac{R_e^2}{R_i^2} - 1 \right) + \frac{\omega R_e^3}{Q_{nom}} \frac{\pi h}{(R_e - R_i)} \frac{R_i}{R_e} \left\{ \frac{R_e}{R_i} - 1 - Ln\left(\frac{R_e}{R_i} \right) \right\}$$

The outline of six rotor blades is shown in solid lines in Figure 2.15. In reality, we have to imagine an infinity of blades of the same shape, very close to each other. We superimpose in dotted lines on the figure the plot of the "neutral" blade calculated in question 2 for the same flow rate. The relative position of the dotted line curve compared to the blades in solid lines makes it possible to apprehend that the machine functions like a pump. The dotted curve is the trajectory that the fluid particles would naturally follow (in the rotating frame) if the rotor had no impact to deviate their trajectory. The movement of the fluid particles, in the rotating frame, goes against the rotational movement of the blades. The fluid is therefore pushed back by the blades. In turbine operation, the fluid particles push the blades according to the direction of rotation of the rotor.

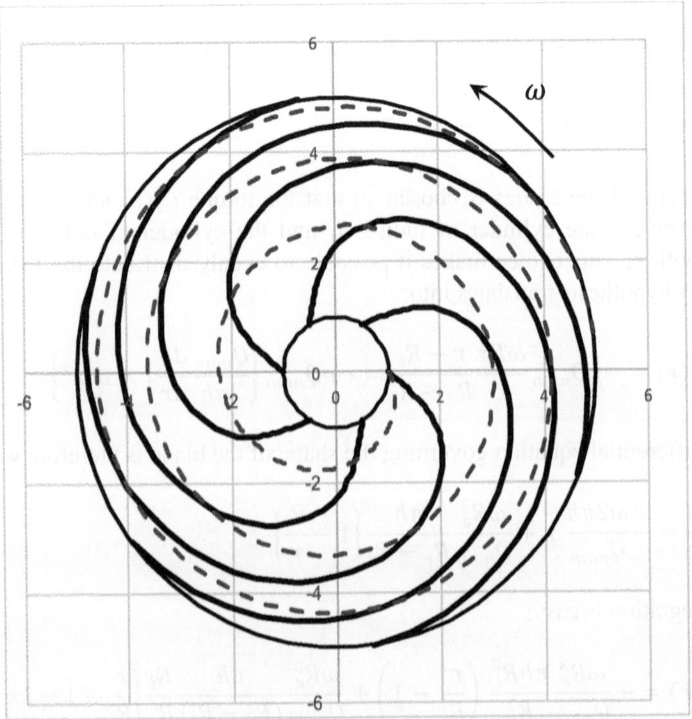

Figure 2.15. *Outline of six rotor blades (black)*

8) The speed components are written, with the results of questions 1, 2 and 5, as:

$$u_r(r) = u'_r(r) = \frac{Q}{2\pi b}\frac{1}{r}$$

$$u_\theta(r) = u'_\theta(r) + \omega r = \frac{Q}{2\pi h}\frac{d\theta}{dr} + \omega r$$

$$u_\theta(R_i) = 0$$

$$u_\theta(R_e) = \frac{\omega R_e}{2}$$

As the kinematics of the flow is of the form $u_r(r)$, $u_\theta(r)$ and $u_z = 0$, Euler's equations simplify to:

$$\rho\left\{u_r\frac{\partial u_r}{\partial r} - \frac{u_\theta^2}{r}\right\} = -\frac{\partial p}{\partial r}$$

$$\rho \frac{u_r}{r} \frac{\partial (r u_\theta)}{\partial r} = -\frac{1}{r} \frac{\partial p}{\partial \theta}$$

$$0 = -\frac{\partial p}{\partial z}$$

In the equation according to the radial direction (first equation), we note the term proportional to u_θ^2. This term, called centrifugal force, plays a role to which we will return later.

According to the azimuthal direction (second equation), the derivative $\partial p/d\theta$ cannot be zero because the angular momentum $r u_\theta$ varies with r. As the flow is axisymmetric, $\partial p/d\theta$ does not depend on θ but only on r. This is only possible because we cross blades when we go through a circle of radius r. The value $\partial p/d\theta \neq 0$ causes the flow to apply a force to each blade, oriented according to the vector $\vec{e}_\theta$. The force applied to a blade tends to zero, as the number of blades is infinite, but the force applied to the rotor is finite and not zero.

It is not possible to calculate the variation $p(r)$ by integrating the Euler equation along the radial direction, because blades are crossed while moving along the radial direction. Integration is to be performed by following a streamline so as to remain in the fluid domain. We therefore evaluate:

$$\frac{\partial p}{\partial r} + \frac{\partial p}{\partial \theta} \frac{d\theta}{dr} = -\rho \left\{ u_r \frac{\partial u_r}{\partial r} - \frac{u_\theta^2}{r} \right\} - \rho u_r \frac{\partial (r u_\theta)}{\partial r} \frac{d\theta}{dr}$$

with the speed components:

$$u_r(r) = \frac{Q}{2\pi h} \frac{1}{r}$$

$$u_\theta(r) = \frac{Q}{2\pi h} \frac{d\theta}{dr} + \omega r$$

Thus:

$$\frac{\partial p}{\partial r} + \frac{\partial p}{\partial \theta} \frac{d\theta}{dr} = \rho \left\{ \left(\frac{Q}{2\pi h} \right)^2 \frac{1}{r^3} + \frac{1}{r} \left\{ \frac{Q}{2\pi h} \frac{d\theta}{dr} + \omega r \right\}^2 \right.$$

$$\left. - \frac{Q}{2\pi h} \frac{1}{r} \frac{\partial}{\partial r} \left\{ \frac{Q}{2\pi h} r \frac{d\theta}{dr} + \omega r^2 \right\} \frac{d\theta}{dr} \right\}$$

By expanding the term to the right of the equality, simplifications appear:

$$\frac{\partial p}{\partial r}+\frac{\partial p}{\partial \theta}\frac{d\theta}{dr}=\rho\left\{\left(\frac{Q}{2\pi h}\right)^2\frac{1}{r^3}+\omega^2 r-\left(\frac{Q}{2\pi h}\right)^2\frac{d^2\theta}{dr^2}\frac{d\theta}{dr}\right\}$$

Integration between $r = R_i$ and $r = R_e$ is easy:

$$p(R_e)-p(R_i)$$
$$=\rho\left\{-\frac{1}{2}\left(\frac{Q}{2\pi h}\right)^2\left\{\frac{1}{R_e^2}-\frac{1}{R_i^2}\right\}+\frac{1}{2}\omega^2\{R_e^2-R_i^2\}\right.$$
$$\left.-\frac{1}{2}\left(\frac{Q}{2\pi h}\right)^2\left\{\left(\frac{d\theta}{dr}\right)^2(R_e)-\left(\frac{d\theta}{dr}\right)^2(R_i)\right\}\right\}$$

One recognizes in the terms on the right of the equality the components of speeds in the moving frame:

$$p(R_e)-p(R_i)=-\frac{1}{2}\rho\{u'^2_r(R_e)-u'^2_r(R_i)\}$$
$$+\frac{1}{2}\rho\omega^2\{R_e^2-R_i^2\}-\frac{1}{2}\rho\{u'^2_\theta(R_e)-u'^2_\theta(R_i)\}$$

For zero flow rate, but with the rotor rotating, the previous expression reduces to:

$$p(R_e)-p(R_i)=\frac{1}{2}\rho\omega^2\{R_e^2-R_i^2\}$$

The pressure on the outer edge of the rotor is greater than the pressure on the inner edge of the rotor because of the centrifugal acceleration due to rotation. The fluid contained inside the channels of the rotor rotates and undergoes this increase in pressure.

If we express the variation of the head by using the components of speed in the relative reference frame, we obtain that the variation of head between the input and the output of the rotor is:

$$\left\{p(R_e)+\frac{1}{2}\rho u'^2_r(R_e)+\frac{1}{2}\rho u'^2_\theta(R_e)\right\}-\left\{p(R_i)+\frac{1}{2}\rho u'^2_r(R_i)+\frac{1}{2}\rho u'^2_\theta(R_i)\right\}$$
$$=\frac{1}{2}\rho\omega^2\{R_e^2-R_i^2\}$$

We recognize here the formulation of Bernoulli's theorem in a rotating reference frame (section 1.2). The rotation of the fluid with the rotor also produces a pressure increase between the inlet and the outlet of the rotor caused by the centrifugal force.

By expressing the speeds in the relative reference as a function of the speeds in the fixed reference, the variation of the head is calculated using the speed components in the fixed reference. We obtain the successive calculation steps:

$$\left\{p(R_e)+\frac{1}{2}\rho u_r^2(R_e)+\frac{1}{2}\rho(u_\theta(R_e)-\omega R_e)^2\right\} \\ -\left\{p(R_i)+\frac{1}{2}\rho u_r^2(R_i)+\frac{1}{2}\rho(u_\theta(R_i)-\omega R_i)^2\right\} \\ =\frac{1}{2}\rho\omega^2\{R_e^2-R_i^2\}$$

$$\left\{p(R_e)+\frac{1}{2}\rho u_r^2(R_e)+\frac{1}{2}\rho u_\theta^2(R_e)-\rho\omega R_e u_\theta(R_e)\right\} \\ -\left\{p(R_i)+\frac{1}{2}\rho u_r^2(R_i)+\frac{1}{2}\rho u_\theta^2(R_i)-\rho\omega R_i u_\theta(R_i)\right\}=0$$

$$\left\{p(R_e)+\frac{1}{2}\rho u_r^2(R_e)+\frac{1}{2}\rho u_\theta^2(R_e)\right\}-\left\{p(R_i)+\frac{1}{2}\rho u_r^2(R_i)+\frac{1}{2}\rho u_\theta^2(R_i)\right\} \\ =\rho\omega R_e u_\theta(R_e)-\rho\omega R_i u_\theta(R_i)=\rho\frac{\omega^2 R_e^2}{2}$$

The variation of the head estimated using the speed components in the fixed frame is therefore equal to the head brought by the rotor to the fluid particles, as obtained in question 5.

These two calculations of the variation of the head between the entrance and the exit of the rotor underline the distinctions between the moving reference frame and the fixed reference frame:

– In the rotating reference, the increase in the head obtained is due to the increase in pressure resulting from the centrifugal force, as when the flow rate is zero. It is an increase in static pressure. In the fixed frame, the increase in the head corresponds to the energy transmitted by the blades to the fluid particles.

– The flows in the fixed frame and in the moving frame are both stationary, but the flow in the fixed frame is *less stationary* than in the moving frame. This means that the flow is strictly stationary in the moving reference frame. On the other hand, in the fixed frame, we always have the same speed at a point when we are in the fluid, but at any point inside the rotor we see the blades scrolling. The process is therefore not stationary, even if we always find the same value of the speed when we are in the fluid domain.

The numerical values of the speed components in the fixed frame and in the moving frame, the kinetic energies of the radial and azimuthal velocity components in the fixed frame and the pressure increase between the inlet and the outlet of the rotor are given in Table 2.4 for the operating conditions of question 6.

r (m)	Q (m³/s)	$u'_r(r)$ (m/s)	$u'_\theta(r)$ (m/s)	$u_r(r)$ (m/s)	$u_\theta(r)$ (m/s)
R_i	0	0	0	0	4.90
R_e	0	0	0	0	24.5
R_i	0.03	6.28	– 4.9	6.28	0
R_e	0.03	1.26	– 12.25	1.26	12.25

Q (m³/s)	$p(R_e) - p(R_i)$ (Pa)	$E_{cin,r}(R_e) - E_{cin,r}(R_i)$ (Pa)	$E_{cin,\theta}(R_e) - E_{cin,\theta}(R_i)$ (Pa)
0	2.88×10^5	0	0
0.03	2.44×10^5	-1.895×10^4	7.50×10^4

Table 2.4. *Numerical applications*

Two remarkable properties should be underlined:

– The increase in the head in the rotor is essentially due to the increase in pressure produced by the centrifugal force. Calling this pump a centrifugal pump is therefore fully justified.

– As the radial component u_r of the speed decreases in $1/r$, the kinetic energy associated with the radial component of the speed decreases between the entry and the exit of the rotor. The same is not true for the azimuthal component u_θ. There is also an increase in the kinetic energy in the fixed frame between the inlet and the outlet of the rotor but it is clearly less than the increase in pressure.

9) The torque and power of the rotor are not changed by reducing the number of blades, as long as these have the same angle at the leading edge and at the trailing edge and the speed components remain the same at the entry and exit of the rotor. In other words, the energy performances are identical. The derivative $\partial p/d\theta$ will be approximately the same between two rotor blades. The pressure difference between the two faces of a blade then increases as the number of rotor blades is reduced and their spacing increases. When the rotor blades are finite in number, the flow between the rotor blades is no longer axisymmetric. It is decelerated near the face in overpressure while it is accelerated near the face in depression, in application of Bernoulli's theorem, so as to increase the pressure difference between the two faces

of the blade[14]. Four current lines between two successive blades of a six-bladed rotor are plotted in Figure 2.16. On the blade pressure side (intrados) and on the blade suction side (extrados), the streamline follows the shape of the blade. At the entrance and at the exit of the rotor, the current lines drawn have an equal spacing because the speed conditions in modulus and direction are the same on the four current lines. On the other hand, between the inlet and the outlet of the rotor, the neighboring streamlines come together when they are on the extrados side (the fluid accelerates and the pressure decreases), while they move apart when they are on the extrados side (the fluid slows down and the pressure increases). Although the flow is modified between the blades, the model with an infinite number of blades remains relevant for estimating the energy performance of the rotor, insofar as the conditions at the entrance and exit of the rotor are identical.

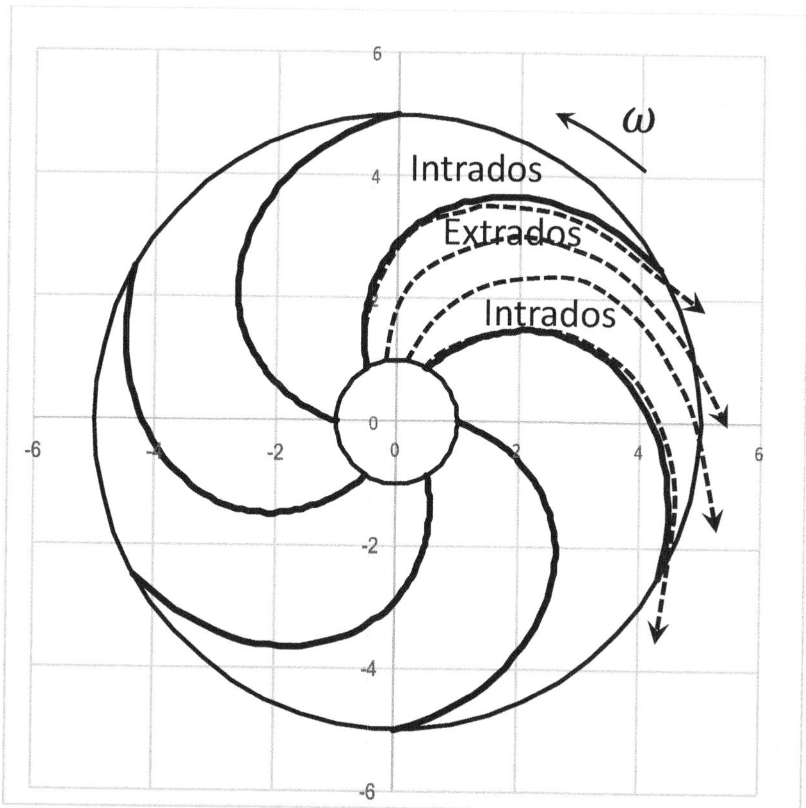

Figure 2.16. *Streamlines between two blades of a six-bladed rotor*

14 See in section 1.2 the formulation of Bernoulli's theorem in a rotating flow.

10) As the length of the suction pipe is short, the linear head loss in the pipe can be neglected. The head along a streamline (dotted line in Figure 2.14) going from far to infinity (more simply to the free surface – point A) to the inlet of the pump (point C) is conserved. Therefore, the pressure at the inlet of the pump is in application of Bernoulli's theorem:

$$P_{asp} = P_{atm} - \rho\frac{U_{asp}^2}{2} - \rho g Z_{asp}$$

The suction speed is deducted from the flow rate:

$$U_{asp} = \frac{Q_{nom}}{\pi R_i^2}$$

The maximum height of the pump above the surface of the basin, to avoid cavitation, is:

$$Z_{asp} < \frac{P_{atm} - P_v(T)}{\rho g} - \frac{U_{asp}^2}{2g}$$

The numerical results are given in Table 2.5.

Q_{nom} (m³/s)	R_i (m)	U_{asp} (m/s)	$P_v(T)$ (Pa)	P_{atm} (Pa)	Z_{asp} (m)
0.030	0.039	6.28	2,300	10^5	7.95

Table 2.5. *Numerical applications relating to cavitation*

The acceleration of the fluid in the suction pipe lowers the pressure at the inlet of the pump. The risk of cavitation appears if the diameter of the pipe is too small or if the pump is located too high above the free surface of the basin.

In the model with an infinite number of blades, cavitation inception occurs at the inlet of the rotor. When the rotor has a finite number of blades, it can appear first on the extrados of the blades (face of the blades in depression).

11) The mass balance on a volute element between the positions θ and $\theta + d\theta$ is written, if the azimuthal component of the speed $u_\theta(R_e)$ is preserved in the volute:

$$u_\theta(R_e)\, h\, H(\theta + d\theta) - u_\theta(R_e)\, h\, H(\theta) = u_r(R_e)\, h\, R_e d\theta$$

We deduce from the passage to the limit $d\theta \to 0$ that:

$$\frac{dH}{d\theta} = \frac{u_r(R_e)}{u_\theta(R_e)} R_e$$

The width at the outlet of the volute is then:

$$H_{s,volute} = \frac{u_r(R_e)}{u_\theta(R_e)} R_e 2\pi$$

2.5. Exercise: radial flow turbine model

This exercise simplifies the operation of a Francis turbine to model it without resorting to fluid mechanics simulation software. The diagram of the turbine is shown in Figure 2.17. The axisymmetric geometry is represented in a cylindrical coordinate system whose Oz axis coincides with the axis of rotation of the rotor. The stator has an infinite number of blades whose trailing edge makes an angle α with the unit vector $\vec{e}_r$. The rotor is a rotor with an infinite number of blades, one of which is drawn in the figure. It is a portion of a circular cylinder of radius R_{bl}, centered at C. Its height (in the direction z) is b and the angle of the arc of a circle on the cylinder is β.

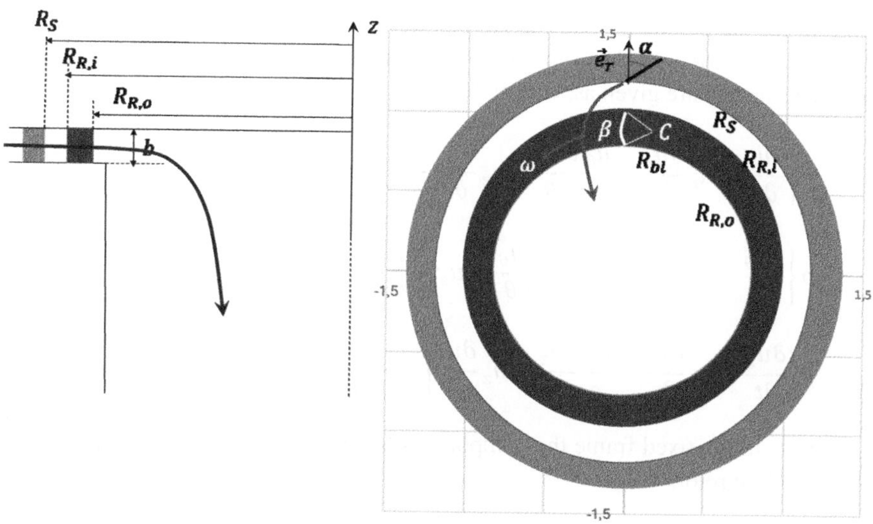

Figure 2.17. *Radial flow turbine geometry. The stator is in the gray light domain and the rotor in the dark gray domain*

The simplification of the model, compared to a Francis turbine, lies in that the stator and the rotor are both placed between two planes perpendicular to the direction Oz, a constant height apart b. Because of this, and because the stator and the rotor have an infinite number of blades, the dependence of the speed components on the coordinates (r, θ, z) is the following:

$$u_r(r)\ ;\ u_\theta(r)\ ;\ u_z = 0$$

The flow is axisymmetric (no dependence with θ) and two-dimensional (no dependence with z) in the channel containing the rotor. The radius at the exit of the stator is denoted by R_S, while the radii at the entrance and at the exit of the rotor are denoted by $R_{R,i}$ and $R_{R,o}$, respectively.

The turbine is implemented using the water resource of a basin whose water surface is located at an altitude h above the turbine. The effective height that can be recovered by the turbine is reduced to $h_{eff} < h$ due to the linear head loss in the penstock joining the upstream reservoir and the turbine.

1) Assuming ideal operation (no head loss other than that recovered by the turbine), write the relationship linking the recovered power $\wp$, the effective height h_{eff} and the flow rate Q passing through the turbine.

2) Express in the fixed frame the components of speed $u_r(R_S)$ and $u_\theta(R_S)$ at the output of the stator. Justify that the angular momentum is conserved in the domain between the stator and the rotor. Euler's equations in a non-rotating cylindrical coordinate system are given as, with p denoting the piezometric pressure:

$$\rho\left\{\frac{\partial u_r}{\partial t} + u_r\frac{\partial u_r}{\partial r} + \frac{u_\theta}{r}\frac{\partial u_r}{\partial \theta} + u_z\frac{\partial u_r}{\partial z} - \frac{u_\theta^2}{r}\right\} = -\frac{\partial p}{\partial r}$$

$$\rho\left\{\frac{\partial u_\theta}{\partial t} + \frac{u_r}{r}\frac{\partial (r u_\theta)}{\partial r} + \frac{u_\theta}{r}\frac{\partial u_\theta}{\partial \theta} + u_z\frac{\partial u_\theta}{\partial z}\right\} = -\frac{1}{r}\frac{\partial p}{\partial \theta}$$

$$\rho\left\{\frac{\partial u_z}{\partial t} + u_r\frac{\partial u_z}{\partial r} + \frac{u_\theta}{r}\frac{\partial u_z}{\partial \theta} + u_z\frac{\partial u_z}{\partial z}\right\} = -\frac{\partial p}{\partial z}$$

Express in the fixed frame the components of speed $u_r(R_{R,i})$ and $u_\theta(R_{R,i})$ at the entrance of the rotor.

Write the expression for the pressures $p(R_S)$ and $p(R_{R,i})$ at the stator outlet and at the rotor inlet, respectively. What should you pay attention to when choosing the angle α and the radii R_S and $R_{R,i}$?

The angular speed of rotation of the rotor is noted ω. Express the speed components $u'_r(R_{R,i})$ and $u'_\theta(R_{R,i})$ at the entrance of the rotor in the cylindrical coordinate system rotating with the rotor.

3) Indicate which condition on the flow allows optimal operation of the turbine.

It is recalled that the rotor has an infinite number of blades. Apply the angular momentum theorem to calculate the torque C_z applied by the flow on the rotor, then the power $\wp$ recovered by the turbine. In which coordinate system (rotating or non-rotating) should this theorem be applied?

4) In this question, we carry out the numerical applications for different stators and rotors contained in the same envelope whose dimensions are $R_S = 1.2$ m, $R_{R,i} = 1.1$ m, $R_{R,o} = 0.9$ m and $b = 0.1$ m. The rotor rotates at speed $n = 600$ rpm and it is desired that the turbine recovers power $\wp = 45$ MW. The flow rate Q and the angle α of the stator blades must be adjusted, depending on the value of the effective height h_{eff}.

For the three values $h_{eff} = 300$ m, $h_{eff} = 500$ m and $h_{eff} = 700$ m, calculate the following values: flow rate Q, stator blade angle α, speed components at the stator outlet and at the rotor inlet, pressure at the rotor inlet. Comment on the results.

5) In this question, one determines the radius R_{bl} and the position of the axis C of the portion of circular cylinder constituting a rotor blade. Using the representation of Figure 2.17, we determine the shape of the blade whose leading edge is in $r = R_{R,i}$ for $\theta = \pi/2$.

The blade is drawn in such a way that in the reference frame of the blade, the flow is tangent to the arc of the blade. Determine the unit vector $\vec{n}_i = n_{i,r}\vec{e}_r + n_{i,\theta}\vec{e}_\theta$ perpendicular to the blade at $r = R_{R,i}$.

At the exit of the rotor, the rotational speed component is $u_\theta(R_{R,o}) = 0$. Determine the unit vector $\vec{n}_o = n_{o,r}\vec{e}_r + n_{o,\theta}\vec{e}_\theta$ perpendicular to the blade at $r = R_{R,o}$.

We can also write that $\vec{n}_i = n_{i,r}\vec{j} - n_{i,\theta}\vec{i}$ in the Cartesian coordinate system, since the position of the blade at the inlet of the rotor is at $(r = R_{R,i}, \theta = \pi/2)$. We designate the position of the blade at the outlet of the rotor by the coordinates $(r = R_{R,o}, \theta)$, of which θ is still undetermined. In the Cartesian coordinate system, express the unit vector $\vec{n}_o = n_{o,r}\vec{e}_r + n_{o,\theta}\vec{e}_\theta$ using the functions cosθ and sinθ as well as the unit vectors $(\vec{i}, \vec{j})$.

When the blade is an arc of a circle, the two lines perpendicular to the blade passing through $r = R_{R,i}$ and $r = R_{R,o}$ intersect at point C, which is at the same distance R_{bl} from the arc of a circle. Use this condition to determine the value of θ defining the position of the trailing edge of the blade. Deduce the radius of curvature R_{bl} of the blade and the position of point C.

For the three values of the effective height $h_{eff} = 300$ m, $h_{eff} = 500$ m and $h_{eff} = 700$ m, the radius of curvature R_{bl} and the position of point C will be calculated. Comment on the results based on a graphic representation of the blade.

2.5.1. *Solution*

1) In ideal operation, the power available is equal to the power recovered by the turbine:

$$\wp = \rho g h_{eff} Q = \Delta H_{turbine} Q$$

2) At the output of the stator, in the fixed mark:

$$u_r(R_S) = -\frac{Q}{2\pi b R_S}$$

$$u_\theta(R_S) = \frac{Q}{2\pi b R_S} \tan\alpha$$

$u_r(R_S)$ is determined using the definition of throughput. The Joukovski condition at the trailing edge of the stator blades is used to express $u_\theta(R_S)$.

The flow is axisymmetric because the stator and the rotor have an infinite number of blades (infinite-bladed rotor). It is independent of altitude z because the rotor and stator are placed between two planes of constant spacing b. Hence, the kinematics:

$$u_r(r)\ ;\ u_\theta(r)\ ;\ u_z = 0$$

Euler's equations in the space between the stator output and the rotor input $(R_{R,i} < r < R_S)$ reduce to:

$$\rho\left\{u_r \frac{\partial u_r}{\partial r} - \frac{u_\theta^2}{r}\right\} = -\frac{\partial p}{\partial r}$$

$$\rho\frac{u_r}{r}\frac{\partial(ru_\theta)}{\partial r}=-\frac{1}{r}\frac{\partial p}{\partial\theta}$$

$$0=-\frac{\partial p}{\partial z}$$

The term on the left of the equality in the Euler equation according to the direction $\vec{e}_\theta$ (second equation) only depends on r. The derivative $\partial p/d\theta$ therefore also depends only on r. We necessarily have:

$$\frac{\partial p}{\partial\theta}=0$$

If $\partial p/d\theta\neq 0$, then $p(r,\theta+2\pi)\neq p(r,\theta)$, which is impossible in the area between the stator and the rotor where there is no solid wall. Consequently, the pressure is also axisymmetric in the domain between the stator and the rotor and the Euler equation according to the direction $\vec{e}_\theta$ reduces to:

$$\rho\frac{u_r}{r}\frac{\partial(ru_\theta)}{\partial r}=0$$

The angular momentum $r\,u_\theta(r)$ is therefore constant and the azimuthal speed varies in the domain between the stator and the rotor ($R_{R,i}<r<R_s$) according to:

$$u_\theta(r)=\frac{A}{r}$$

The conservation of flow rate ($Q=-2\pi r\,b\,u_r(r)$) determines the radial component of the speed:

$$u_r(r)=-\frac{Q}{2\pi b}\frac{1}{r}$$

Consequently, the speed components in the fixed frame at the entrance of the rotor are given as:

$$u_r(R_{R,i})=-\frac{Q}{2\pi bR_{R,i}}$$

$$u_\theta(R_{R,i})=\frac{Q}{2\pi bR_{R,i}}\tan\alpha$$

Since the fluid particles do not undergo a head loss between the stator outlet and the rotor inlet, Bernoulli's theorem makes it possible to express the piezometric pressure at the outlet of the stator and at the inlet of the rotor:

$$p(R_S) = P_{atm} + \rho g h_{eff} - \rho \frac{1}{2}\left(\frac{Q}{2\pi b}\right)^2 \frac{1}{R_S^2}(1 + \tan^2\alpha)$$

$$p(R_{R,i}) = P_{atm} + \rho g h_{eff} - \rho \frac{1}{2}\left(\frac{Q}{2\pi b}\right)^2 \frac{1}{R_{R,i}^2}(1 + \tan^2\alpha)$$

The linear head loss in the penstock has been taken into account by replacing the actual height h with the effective height h_{eff}. The pressure must remain higher than the vapor pressure ($P_v = 2{,}300$ Pa at the temperature of 20°C) so as to avoid cavitation. The stator converts pressure energy into kinetic energy. If cavitation were to occur, it would be triggered first at the inlet of the rotor, since the speed is greatest there.

Cavitation is avoided if:

$$p(R_{R,i}) = P_{atm} + \rho g h_{eff} - \rho \frac{1}{2}\left(\frac{Q}{2\pi b R_{R,i}}\right)^2 \frac{1}{\cos^2\alpha} > P_v$$

either by checking:

$$\cos^2\alpha > \left(\frac{Q}{2\pi b R_{R,i}}\right)^2 \frac{1}{\frac{2(P_{atm} - P_v)}{\rho} + 2g h_{eff}}$$

The angle α of the trailing edge of the stator blades must therefore verify:

$$\alpha < \alpha_{max} = \text{Arccos}\left(\frac{\frac{Q}{2\pi b R_{R,i}}}{\sqrt{\frac{2(P_{atm} - P_v)}{\rho} + 2g h_{eff}}}\right)$$

At the entrance to the rotor, the speed components in the rotating frame are given as:

$$u'_r(R_{R,i}) = -\frac{Q}{2\pi b R_{R,i}}$$

$$u'_\theta(R_{R,i}) = \frac{Q}{2\pi b R_{R,i}}\tan\alpha - \omega R_{R,i}$$

3) It is desirable that the rotational component of the flow u_θ be zero at the outlet of the rotor, that is:

$$u_\theta\left(R_{R,o}\right) = 0$$

If this is not the case, the kinetic energy of rotation cannot be recovered (unless the machine is fitted with a downstream stator) and it would be dissipated in a head loss downstream.

The angular momentum theorem (section 1.9) must be applied in a frame where the flow is stationary. It is so exactly in the rotating mobile frame. The torque is then deduced from:

$$C_{z,rotor} = -\iint_{S_d} (\rho r u'_\theta)\left(\vec{u'}.\vec{n}\right)ds - \iiint_V \rho 2\omega r u'_r dv$$

The speed components with a prime are those in the rotating frame. S_d denotes the delivery sections at the inlet (circular cylinder of radius $R_{R,i}$ and height b) and at the outlet (circular cylinder of radius $R_{R,o}$ and height b) of the rotor.

In the particular geometry of the flow (constant channel height), the speed:

$$u'_r(r) = u_r(r) = -\frac{Q}{2\pi b r}$$

is known and the integral of the Coriolis force over the volume can be calculated.

For a rotor with an infinite number of blades, it is also possible to consider that the flow is steady in the fixed frame, and the angular momentum theorem can alternatively be written as:

$$C_{z,rotor} = -\iint_{S_d} (\rho r u_\theta)\left(\vec{u}.\vec{n}\right)ds$$

If the rotational component of the flow is removed at the rotor outlet $\left(u_\theta\left(R_{R,o}\right) = 0\right)$, the integral over the discharge surface need only be calculated in the rotor inlet section, hence:

$$C_{z,rotor} = -\iint_{S_d} (\rho r u_\theta)\left(\vec{u}.\vec{n}\right)ds = \rho Q \frac{Q}{2\pi b} \tan\alpha$$

Subsequently:

$$\wp = C_{z,rotor}\omega = \omega\rho Q \frac{Q}{2\pi b}\tan\alpha = \Delta H_{turbine} Q = \rho g h_{eff} Q$$

The angle of the blades of the stator and the angular speed of rotation of the rotor are therefore linked by the relationship:

$$\tan\alpha = \frac{g h_{eff} 2\pi b}{\omega Q}$$

4) The numerical results are shown in Table 2.6.

h_{eff} (m)	Q (m³/s)	$\tan\alpha$ (–)	α (°)	α_{max} (°)
300	15.29	1.926	62.6	73.52
500	9.17	5.349	79.4	82.37
700	6.55	10.484	84.6	85.39

h_{eff} (m)	Q (m³/s)	$u_r(R_S)$ (m/s)	$u_\theta(R_S)$ (m/s)	$u_r(R_{R,i})$ (m/s)	$u_\theta(R_{R,i})$ (m/s)	$p(R_{R,i})$ (Pa)
300	15.29	– 20.3	39.1	– 22.1	42.6	1.89×10^6
500	9.17	– 12.2	65.1	– 13.3	71.0	2.40×10^6
700	6.55	– 8.7	91.1	– 9.5	99.4	1.98×10^6

Table 2.6. *Numerical values of the dimensioning of a 45 MW power turbine, according to the effective height*

The lower the effective head, the greater the flow rate must be to provide the same power.

The greater the effective height, the greater the angle of the stator blades and the greater the rotational kinetic energy at the output of the stator. The stator converts pressure energy into rotational kinetic energy. For the chosen conditions, cavitation is avoided at the rotor inlet as indicated by the pressure values. However, the greater the effective height, the more precise the adjustment of the stator angle must be. For effective height h_{eff} = 700 m, a 1° increase in stator angle would suffice to cause cavitation.

5) Considering the speed components at the entrance of the rotor in the rotating frame:

$$u'_r(R_{R,i}) = -\frac{Q}{2\pi b R_{R,i}}$$

$$u'_\theta(R_{R,i}) = \frac{Q}{2\pi b R_{R,i}}\tan\alpha - \omega R_{R,i}$$

the vector normal to the blade at the rotor inlet is parallel to:

$$\left\{\tan\alpha - \frac{\omega 2\pi b R_{R,i}^2}{Q}\right\}\vec{e}_r + \vec{e}_\theta = \left\{\tan\alpha - \frac{gh_{eff}(2\pi b)^2 R_{R,i}^2}{Q^2 \tan\alpha}\right\}\vec{e}_r + \vec{e}_\theta$$
$$= \tan\alpha(1-B)\vec{e}_r + \vec{e}_\theta$$

where the angular speed of rotation has been replaced according to the relationship established at the end of question 3. In this expression:

$$B = gh_{eff}\left(\frac{2\pi b R_{R,i}}{Q\tan\alpha}\right)^2$$

The unit vector perpendicular to the leading edge of the blade (Figure 2.18) is therefore $n_{i,r}\vec{e}_r + n_{i,\theta}\vec{e}_\theta$, with:

$$n_{i,r} = -t_{i,\theta} = -\frac{(1-B)\tan\alpha}{\sqrt{1+(1-B)^2\tan^2\alpha}}$$

$$n_{i,\theta} = t_{i,r} = -\frac{1}{\sqrt{1+(1-B)^2\tan^2\alpha}}$$

We note that $t_{i,\theta} > 0$ when $B < 1$, while $t_{i,\theta} < 0$ when $B > 1$.

At the exit of the rotor, the speed components in the rotating frame are given as:

$$u'_r(R_{R,o}) = -\frac{Q}{2\pi b R_{R,o}}$$

$$u'_\theta(R_{R,o}) = -\omega R_{R,o}$$

The last relation ensures that the rotation component is suppressed at the output of the rotor ($u_\theta(R_{R,o}) = 0$).

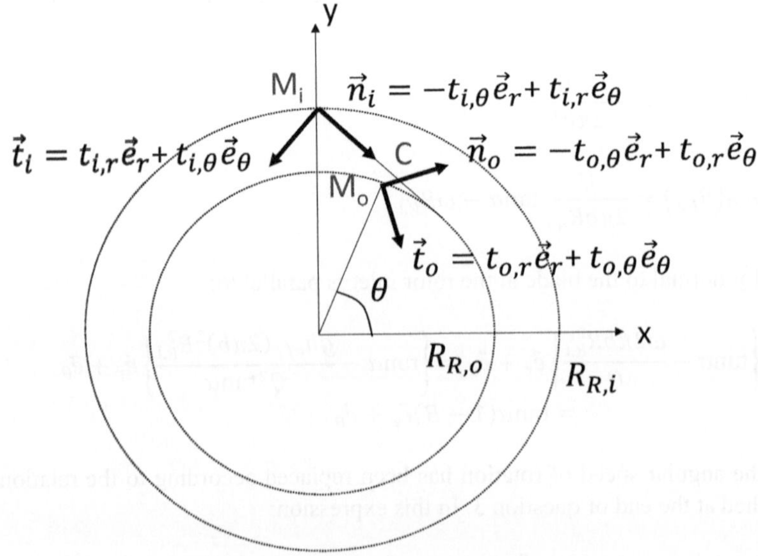

Figure 2.18. *Unit vectors tangent and perpendicular to both ends of the blade*

The vector normal to the blade at the exit of the rotor is parallel to:

$$\frac{\omega 2\pi b R_{R,o}^2}{Q}\vec{e}_r - \vec{e}_\theta = \frac{gh_{eff}(2\pi b)^2 R_{R,o}^2}{Q^2 \tan\alpha}\vec{e}_r - \vec{e}_\theta = B\tan\alpha\left(\frac{R_{R,o}}{R_{R,i}}\right)^2 \vec{e}_r - \vec{e}_\theta$$

The unit vector normal to the blade $n_{o,r}\vec{e}_r + n_{o,\theta}\vec{e}_\theta$ at the trailing edge is therefore:

$$n_{o,r} = -t_{o,\theta} = \frac{B\tan\alpha\left(\frac{R_{R,o}}{R_{R,i}}\right)^2}{\sqrt{1 + B^2\left(\frac{R_{R,o}}{R_{R,i}}\right)^4 \tan^2\alpha}}$$

$$n_{o,\theta} = t_{o,r} = -\frac{1}{\sqrt{1 + B^2\left(\frac{R_{R,o}}{R_{R,i}}\right)^4 \tan^2\alpha}}$$

The tangent and normal unit vectors at both ends of the blade are shown in Figure 2.18.

The position M_i of the blade at the inlet of the rotor $(r = R_{R,i}, \theta = \pi/2)$ is such that in the Cartesian coordinate system, $\vec{n}_i = n_{i,r}\vec{e}_r + n_{i,\theta}\vec{e}_\theta = n_{i,r}\vec{j} - n_{i,\theta}\vec{\imath}$.

By locating the position of the point M_o of the blade at the exit of the rotor by its coordinates $(r = R_{R,o}, \theta)$, where θ is undetermined at this stage, then:

$$\begin{aligned}\vec{n}_o = n_{o,r}\vec{e}_r + n_{o,\theta}\vec{e}_\theta &= n_{o,r}\{\cos\theta\vec{\imath} + \sin\theta\vec{j}\} + n_{o,\theta}\{-\sin\theta\vec{\imath} + \cos\theta\vec{j}\} \\ &= \{n_{o,r}\cos\theta - n_{o,\theta}\sin\theta\}\vec{\imath} + \{n_{o,\theta}\cos\theta + n_{o,r}\sin\theta\}\vec{j}\end{aligned}$$

The lines perpendicular to the blade at both ends intersect at point C, which is defined as the solution of:

$$\begin{aligned}\overrightarrow{OC} = \overrightarrow{OM_\iota} + \overrightarrow{M_\iota C} = R_{R,i}\vec{j} + l_i\vec{n}_i &= R_{R,i}\vec{j} + l_i\{n_{i,r}\vec{j} - n_{i,\theta}\vec{\imath}\} \\ &= -l_i n_{i,\theta}\vec{\imath} + \{R_{R,i} + l_i n_{i,r}\}\vec{j}\end{aligned}$$

and:

$$\begin{aligned}\overrightarrow{OC} = \overrightarrow{OM_o} + \overrightarrow{M_o C} &= R_{R,o}\{\cos\theta\vec{\imath} + \sin\theta\vec{j}\} + l_o\vec{n}_o \\ &= R_{R,o}\{\cos\theta\vec{\imath} + \sin\theta\vec{j}\} + l_o\{n_{o,r}cos\theta - n_{o,\theta}sin\theta\}\vec{\imath} \\ &\quad + l_o\{n_{o,\theta}\cos\theta + n_{o,r}\sin\theta\}\vec{j} \\ &= \{R_{R,o}\cos\theta + l_o(n_{o,r}\cos\theta - n_{o,\theta}\sin\theta)\}\vec{\imath} \\ &\quad + \{R_{R,o}\sin\theta + l_o(n_{o,\theta}\cos\theta + n_{o,r}\sin\theta)\}\vec{j}\end{aligned}$$

By comparing the two expressions of $\overrightarrow{OC}$, the two following relations must be respected, respectively, according to the vectors $\vec{\imath}$ and $\vec{j}$:

$$R_{R,o}\cos\theta + l_o(n_{o,r}\cos\theta - n_{o,\theta}\sin\theta) = -l_i n_{i,\theta}$$

$$R_{R,o}\sin\theta + l_o(n_{o,\theta}\cos\theta + n_{o,r}\sin\theta) = R_{R,i} + l_i n_{i,r}$$

The variations of l_i and l_o with θ are calculated from this linear system. When the blade is an arc of a circle, the angle θ determining the position M_o of the blade at the outlet of the rotor is determined by choosing the value of θ such that $|l_o| = |l_i|$.

The radius of the blade is then:

$$|l_i| = |l_o| = \left|\frac{R_{R,o}\cos\theta}{n_{o,\theta}\sin\theta - n_{o,r}\cos\theta - n_{i,\theta}}\right| = \left|\frac{R_{R,i} - R_{R,o}\sin\theta}{n_{o,\theta}\cos\theta + n_{o,r}\sin\theta - n_{i,r}}\right|$$

and the position of point C is:

$$\overrightarrow{OC} = -l_i n_{i,\theta} \vec{\imath} + \{R_{R,i} + l_i n_{i,r}\} \vec{\jmath}$$

Figure 2.19 shows the shape of a blade (in bold solid line) for three values of the effective height. The thin gray lines connect the ends of the blade to point C. The segments in thin dotted lines extending from the blade indicate the direction of the speed, thus verifying that the flow is tangent to the two ends of the blade.

While for the lowest effective height $h_{eff} = 300$ m, the blade is almost a flat plate, its curvature increases as the effective height increases. For the greatest effective height $h_{eff} = 700$ m, the shape of the blade resembles that of a Pelton turbine bucket.

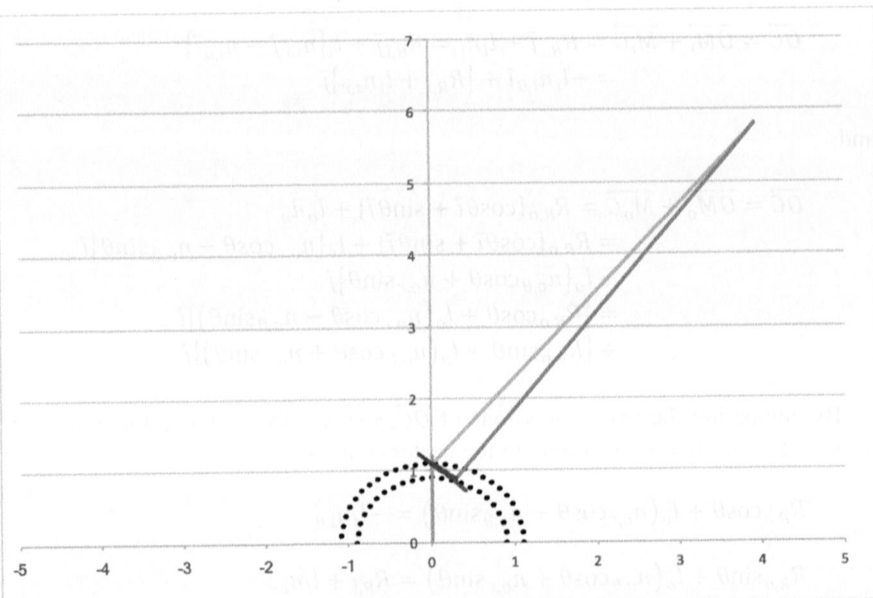

$h_{eff} = 300$ m

Figure 2.19A. *Shape of the blade (in bold solid line) for three values of h_{eff}*

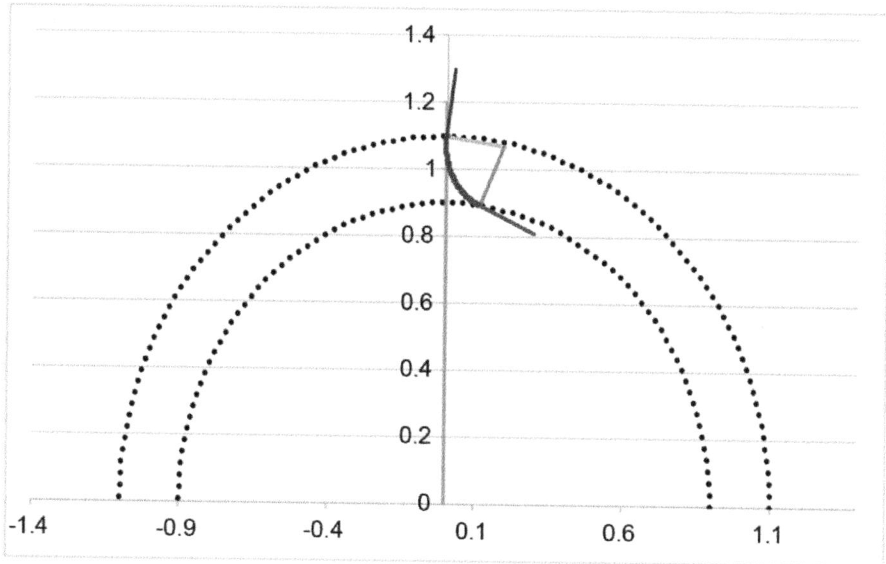

$h_{eff} = 500$ m

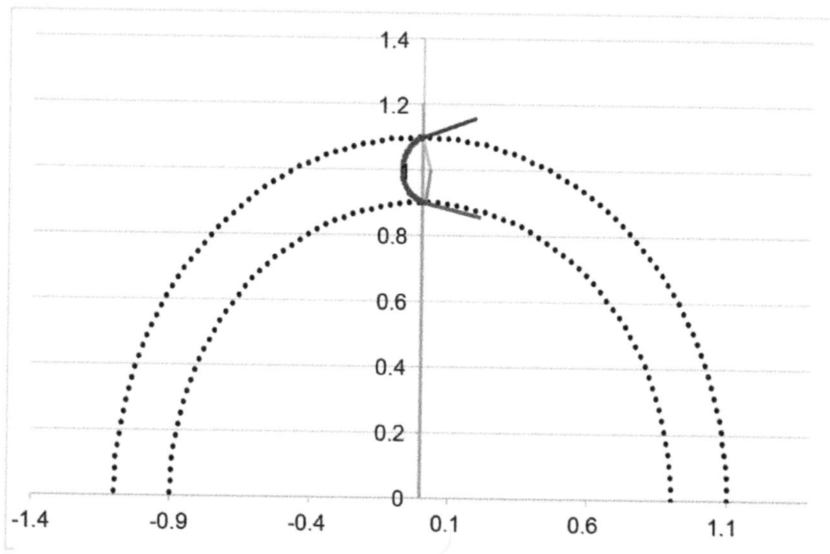

$h_{eff} = 700$ m

Figure 2.19B. *Shape of the blade (in bold solid line) for three values of h_{eff} (continued)*

3

Wind Power

3.1. General

3.1.1. *The wind resource*

The resource of wind turbines is the kinetic energy of the wind. Recordings of wind speed, made over a long period in a territory, make it possible to map the energy resource of the wind. Such a map, shown in Figure 3.1, confirms that the most important wind power is located in the northwest and southeast parts of the French territory.

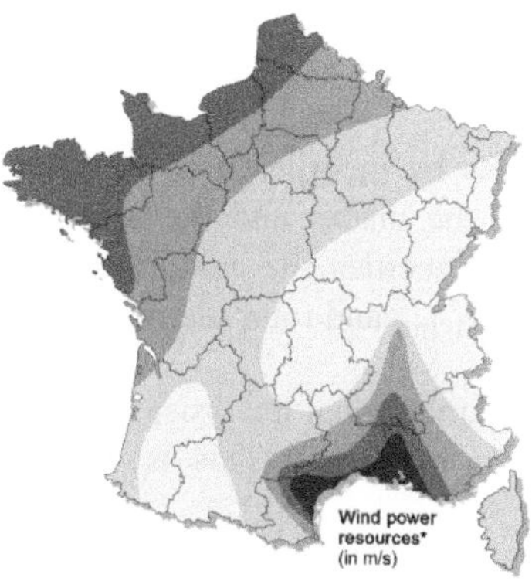

Dense wooded countryside, woodland, suburbs	Open countryside, scattered obstacles	Flat pasture, small bushes	Lakes, sea	Ridges**, hills	
<3.5	<4.5	<5.0	<5.5	<7.0	Zone 1
3.5 - 4.5	4.5 - 5.5	5.0 - 6.0	5.5 - 7.0	7.0 - 8.5	Zone 2
4.5 - 5.0	5.5 - 6.5	6.0 - 7.0	7.0 - 8.0	8.5 - 10.0	Zone 3
5.0 - 6.0	6.5 - 7.5	7.0 - 8.5	8.0 - 9.0	10.0 - 11.5	Zone 4
>6.0	>7.5	>8.5	>9.0	>11.5	Zone 5

* Wind speed at 50 meters above the ground according to the topography
** Mountainous zones require a specific study of the resources

Figure 3.1. *Mapping of the wind resource in the French territory[1]. For a color version of this figure, see www.iste.co.uk/mory/mechanical.zip*

1 Source: ADEME (2011).

The wind is an intermittent resource by nature. For power $\wp$ installed in wind converters, the load factor assesses the degree of intermittency by calculating the ratio of the energy E_{annual} recovered over the course of a year to the installed power multiplied by the duration of a year:

$$f = \frac{E_{annual}}{\wp \times 24 \times 365}$$

This formula requires evaluating power and energy with corresponding units (kW and kWh, MW and MWh, etc.). The load factor estimates the fraction of a year during which the wind turbine can be considered to have operated at nominal power. Similarly, a monthly load factor is defined. Its variation over the course of a year indicates that wind generation is significantly greater during the winter period.

3.1.2. *Energy conversion by a wind turbine*

Two exercises proposed in this chapter (sections 3.2 and 3.3) study the conversion of the kinetic energy of the wind by a wind turbine.

The first exercise demonstrates Betz's famous formula. This formula determines the maximum power that can be harvested by a wind turbine. It is commonly used because it simply evaluates the order of magnitude of the power of a wind turbine for a given wind speed. However, the Betz model does not describe the process of converting the kinetic energy of the wind by the turbine rotor into mechanical power driving the alternator.

A second exercise models this conversion process by incorporating the rotor. The power recovered as a function of the angle of the blades and the speed of rotation of the rotor is thus determined. This model specifies the conditions that maximize the power harvested for a given wind speed U_{wind} and rotor diameter D.

It is interesting to establish beforehand, by using dimensional analysis, the form in which the power recovered by a wind turbine is expressed. This is what is proposed below in the form of an exercise.

3.1.2.1. *Exercise*

Determine, by appealing to the Vashy–Buckingham theorem, the dimensional numbers making it possible to express the relation linking the power $\wp$ recovered by the wind turbine to the speed of the wind U_{wind}, to the diameter D of the rotor, as well as to the other parameters which it seems necessary to take into account.

3.1.2.2. *Solution*

As this is a problem of fluid mechanics, the flow of air in the wind turbine, it seems necessary to involve the density of the air ρ and its kinematic viscosity ν. However, the flow is a large-scale flow and the Reynolds number $U_{wind}D/\nu$ characterizing it is very large. The effect of viscosity can therefore be neglected.

It is important a priori to also involve in this relation the angular speed ω of rotation of the rotor because it can be chosen independently and it probably influences the result.

On the other hand, we will not incorporate any other geometric parameter of the rotor (width L and angle of the blades, etc.), with the exception of its diameter. For a wind turbine with the same geometry, if the size is changed, the blade width must be modified so as to keep the same value of the ratio L/D and the angles remain the same.

The relation must therefore link the parameters listed in Table 3.1 with their dimensions.

Power	**Wind speed**	**Rotor diameter**	**Air density**	**Angular speed of rotation of the rotor**
$\wp$	U_{wind}	D	ρ	ω
$\mathbf{W = kg\ m^2\ s^{-3}}$	$\mathrm{m\ s^{-1}}$	m	$\mathrm{kg\ m^{-3}}$	$\mathrm{s^{-1}}$

Table 3.1. *Operating parameters of a wind turbine with their units*

By virtue of the Vashy–Buckingham theorem, since the relation binds $n = 5$ parameters involving $p = 3$ dimensions (kg, m, s), it can be written as a relation between $n - p = 2$ dimensionless numbers.

As far as possible, we choose to show the power $\wp$ only in the first number (this is the quantity we want to estimate) and the angular speed of rotation ω only in the second number. The unique possible expressions of the two dimensionless numbers are then proportional to:

$$\frac{\wp}{\rho U_{wind}^3 D^2} \quad \text{and} \quad \frac{\omega D}{U_{wind}}$$

These two numbers are related by a relation. Therefore, the harvested power varies with the cube of the wind speed if $\omega D/U_{wind}$ is kept constant.

We usually use the power coefficient:

$$c_p = \frac{\wp}{\frac{1}{2}\rho U_{wind}^3 \pi R^2}$$

to assess the performance of a wind turbine. R is the radius of the wind turbine, (πR^2) the area swept by the blades of the wind turbine, $(\pi R^2 U_{wind})$ being the maximum flow rate that can pass through the wind turbine, while $\frac{1}{2}\rho U_{wind}^2$ is the kinetic energy of the wind per unit volume.

We also use the tip speed ratio:

$$\lambda = \frac{\omega R}{U_{wind}}$$

which is the ratio of the rotor speed at the tip of the blade to the wind speed.

a) b)

Figure 3.2. *Photographs of (a) a Vestas V90 wind turbine and (b) an Openhydro marine current turbine*[2]

2 Source for Figure 3.2(a): https://www.batiproduits.com/fiche/produits/eolienne-pour-sites-a-fai-bles-turbulences-p69095648.html; source for Figure 3.2(b): © EDF – Nicolas Gérard. All rights reserved.

Despite differing geometries and operating conditions, the dimensioning methods studied in this chapter apply to both wind turbines and marine current turbines, two examples of which are shown in Figure 3.2. Data specific to marine current turbines will be discussed in Chapter 4 devoted to tidal energy. In the remainder of this section, we focus on wind energy.

Figure 3.3 shows the variations $c_p(\lambda)$ for different wind turbines. This graph will be discussed in the exercise in section 3.3.

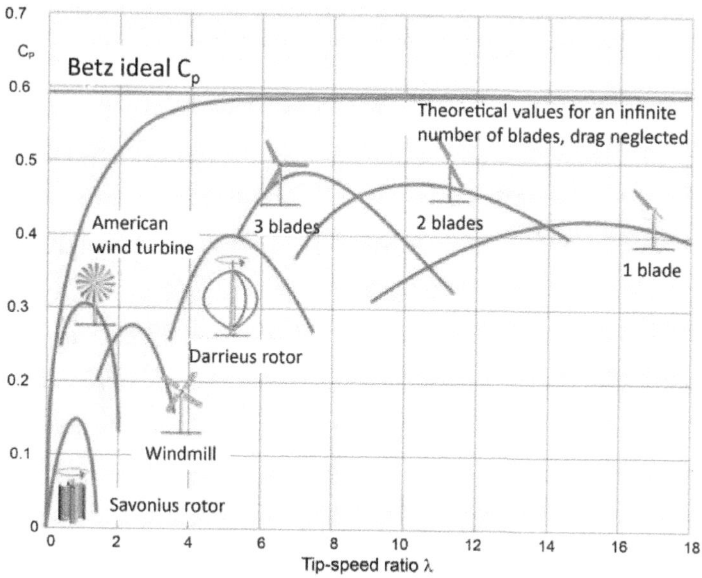

Figure 3.3. *Variation of the power coefficient c_p with the tip speed ratio λ for different wind turbine technologies*[3]

3.1.3. *The development of the wind power sector in France and in Europe*

The development of the onshore wind power sector began in France in the year 2000. The connected power reached 16,930 MW as of June 30, 2020[4]. However, let

3 Available at: www.energieplus-lesite.be/index.php?id=16657.

4 "Panorama de l'électricité renouvelable au 30 juin 2020", publication of the Syndicat des Energies Renouvelables (SER).

us not lose sight of the fact that, although this power is equivalent to that of about 17 nuclear reactors, the wind energy produced is comparable to that of four nuclear reactors, due to the load factor of wind power. Installed power produced 39.1 TWh year-on-year for 2020, which corresponds to 8.6% of French electricity consumption.

The wind power industry is changing rapidly. The reader is referred to the annual reports of the renewable energies syndicate for the most recent and complete figures. We focus here on the salient points. Figure 3.4 shows the breakdown by French region (as of June 30, 2020) of installed capacity, annual production and average load factor. The load factor map can be deduced from the other two.

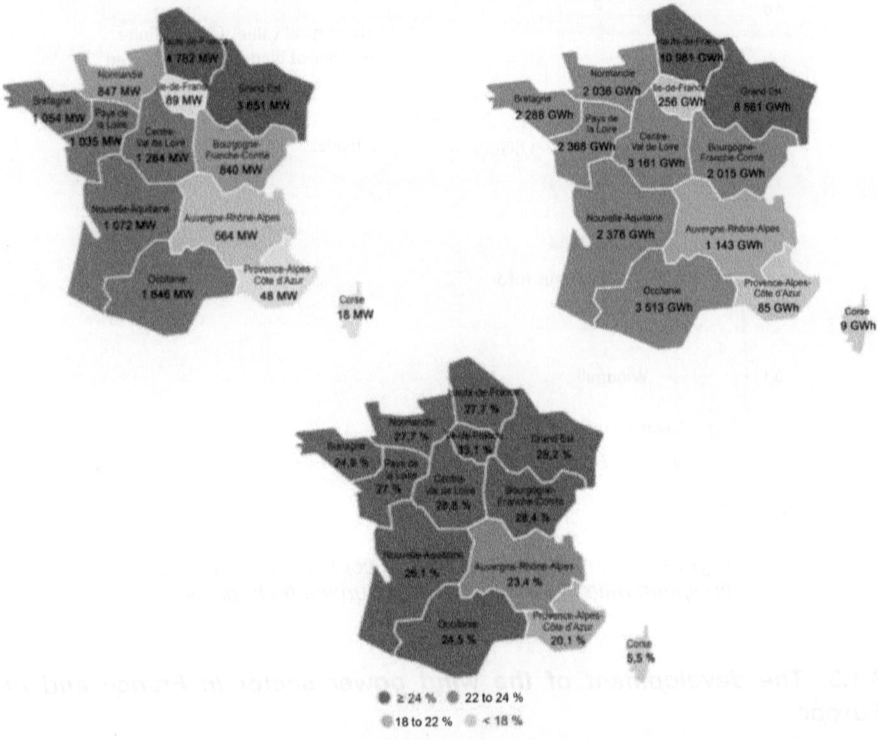

Figure 3.4. *Breakdown (by French region) as of June 30, 2020 (rolling year) of installed capacity, annual production and average wind load factor*[5]*. For a color version of this figure, see www.iste.co.uk/mory/mechanical.zip*

5 "Panorama de l'électricité renouvelable au 30 juin 2020", publication of the Syndicat des Energies Renouvelables (SER).

It is useful to present the three maps simultaneously to bring out some observations:

1) The load factors do not vary significantly from one region to another. They are weakly correlated with the wind resource map (Figure 3.1).

2) Despite a seemingly modest resource (see Figure 3.1), some regions have invested heavily and successfully in the wind power sector, installing significant power and recovering significant production with a significant load factor.

3) These findings do not call into question the wind statistics to decide on the installation of a wind turbine. They question the relevance of an overall decision to invest or not in a territory in the wind power sector on the sole basis of a large-scale map of the wind resource.

The current maps hide the greater disparities that existed before the amalgamation of regions in 2015. The map for 2015 showed an installed capacity of 0 MW for the former Aquitaine. The political choice not to invest in the wind power sector was explained by a study carried out in 2008 at the request of the Regional Council of Aquitaine and ADEME. This map, which refined the resource to a finer scale than that shown in Figure 3.1, was not considered to have sufficient potential to motivate investment in the wind power sector in the territory.

The economic and financial data presented in a recent ADEME (2020) report confirm the maturity of the onshore wind power sector. ADEME indicates an investment cost of between €1,400 and €1,620 per kW installed. For an operating period of 25 years, the levelized cost of energy production (LCOE) is between €50 and €71 per MWh.

Data on installed capacity, energy production, load factor and percentage of electricity consumption of wind power in Europe show significant variations from one country to another. Several countries ensure a significant part of their electricity consumption thanks to wind energy: Denmark (48%), Ireland (33%), Portugal (27%), United Kingdom (22%), Spain (21%), France (7%) (Wikipedia source for the year 2019). Wikipedia also reports a significant supply of offshore wind power in 2019 in the United Kingdom, Germany, Denmark, the Netherlands and Belgium.

In France, President N. Sarkozy set the goal in 2011 of installing 3,000 MW of offshore wind power by 2015. Four wind farms were awarded to EDF and Iberdola in 2012 and two others to Engie in 2014. Each with a capacity of 500 MW, they are to be built offshore from Calvados, Seine-Maritime, Loire-Atlantique, Côtes-d'Armor and Vendée. Another wind farm was awarded in 2019 offshore from Dunkirk to the EDF-Innogy-Enbridge consortium. The realization is slow; their commissioning is now announced between 2022 and 2027. Technological problems related to the anchoring of offshore wind turbine floating platforms and the repatriation of offshore electricity (connectors) seem to be under control today. The difficulty in defining the sites of implantation (conflicts of use linked to fishing, navigation, military and meteorological radars, etc.), administrative delays, legal recourse and the construction of factories to supply the equipment help explain the slow start of offshore wind power in France. The limited data provided by the Syndicat des Energies Renewables on offshore wind power is an indicator of the difficulty of offshore wind power industry to develop in France.

ADEME (2020) indicates a production cost of between €69 and €92 per MWh for installed offshore wind power commissioned in 2017/2018 in Germany, the United Kingdom and the Netherlands. The forecasts put forward for 2050 are for a price in the range of €36–54 per MWh for wind turbines placed at sea and €58–71 per MWh for floating offshore wind turbines.

3.2. Exercise: flow through a wind turbine – Betz formula

The flow around a wind turbine is schematized in Figure 3.5. The air which crosses the surface $S = \pi D^2/4$ swept by the blades of the wind turbine (D designates the diameter of the wind turbine) flows in a flow tube (indicated by a solid line in the figure) limited far upstream by the surface S_1, far downstream by the surface S_2 and whose lateral surface is $S_{L,near}$.

Far from the wind turbine, except in its wake, the flow is not modified by the wind turbine. The speed is therefore taken equal to the speed of the wind U_{wind} on the surfaces S_1, S_{ext} and $S_{L,far}$. In section S_2 in the wake at the rear of the wind turbine, the speed U_{wake} is homogeneous but different from the speed of the wind U_{wind}. On any flowing surface (perpendicular to the axis Ox) inside the flow tube surrounded by the surface, $S_{L,near}$ the velocity is also assumed to be homogeneous.

The Betz model quantifies the maximum power that can be harvested by a wind turbine as a function of the wind speed U_{wind} and the size of the wind turbine quantified by the surface S. It simplifies the problem by ignoring the rotation of the flow induced in the wake by the rotation of the blades. It is considered that the fluid does not rotate around the axis Ox in the section S_2.

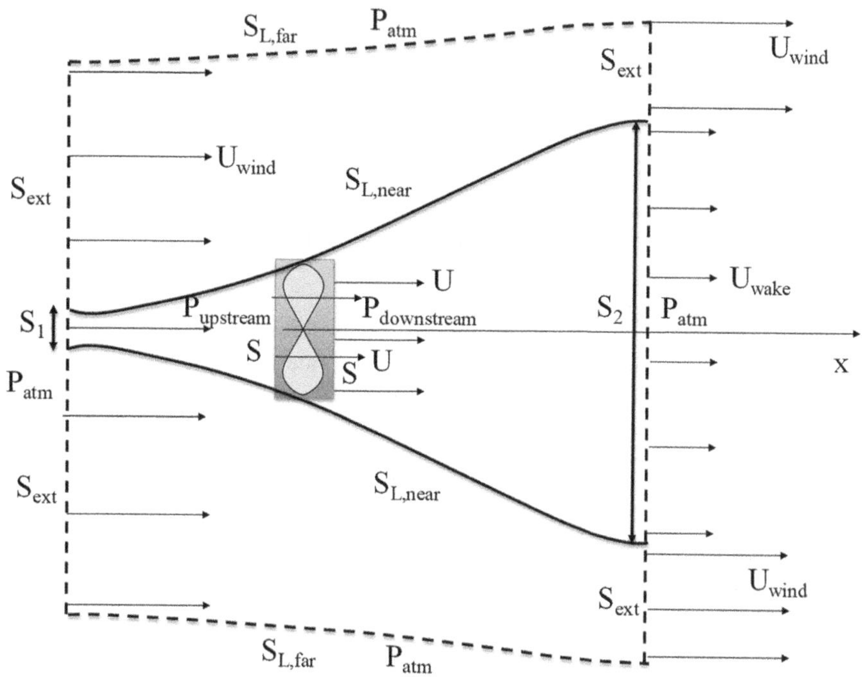

Figure 3.5. *Representation of the flow around a wind turbine*

The fluid particles are slowed down going from upstream to downstream in the flow tube and they give up part of their energy to the blades to turn the rotor. The consequence of this is that the speed U_{wake} in section S_2 behind the wind turbine is different from the wind speed U_{wind}. Far from the wind turbine, the speed is therefore U_{wind}, except on the discharge section S_2 of the wake. Away from the wind turbine, the flow is parallel and therefore the pressure is atmospheric pressure P_{atm}.

1) The Betz model assumes ideal turbine operation: fluid particles passing through the wind turbine give up the same head to the wind turbine rotor and experience no further head loss. Justify that the pressure is hydrostatic in the sections S_1, S_{ext}, $S_{L,far}$ and S_2. For the rest of the exercise, the pressure will designate the piezometric pressure $p + \rho g z$.

2) Using energy reasoning, explain why the wind turbine causes a widening of the flow tube which crosses its rotor, that is, $S_2 > S_1$.

3) The flow rate surfaces near the upstream and near the downstream of the wind turbine are identical and equal to the surface S swept by the wind turbine. The flow rate velocity in these two sections is therefore identical and it is noted U. One neglects in the Betz model any head loss of the fluid particles, which is not recovered by the wind mill. Establish the expressions of the pressures $P_{upstream}$ and $P_{downstream}$ upstream near and downstream near the wind turbine as a function of U_{wind}, U_{wake}, U and ρ. From this, deduce $\Delta H = P_{upstream} - P_{downstream}$. Schematically plot the variations of velocity, pressure and head along a streamline from section S_1 to section S_2.

4) Express the power received by the wind turbine. To make subsequent calculations more readable, we will write the flow rate US. This expression shows two unknowns, U and U_{wake}.

5) This question and the next establish a relation between U_{wind}, U_{wake} and U by two applications of the momentum theorem.

Apply the momentum theorem to the large domain surrounded by the dotted lines, for which the lateral surface $S_{L,far}$ surrounds a flow tube, and express the force exerted on the wind turbine as a function of the speeds U_{wind} and U_{wake} and flow rate US. What is the direction of this force?

6) Apply the momentum theorem to the rectangular domain enclosing the wind turbine with the discharge surfaces S upstream and downstream. By comparison with the result of question 5, determine the speed U. Use this relationship to express the power obtained in question 4 as a function of ρ, U_{wind}, $a = U_{wake}/U_{wind}$ and the section of the wind turbine S.

7) Express the maximum power received by a wind turbine with surface S for a wind speed U_{wind} (Betz's formula). Discuss the two limiting cases between which the wind turbine recovers energy. How could you define the maximum theoretical efficiency of a wind turbine and what would it be worth?

Comment on Betz's formula and discuss its application for a marine current turbine placed at sea in a tidal current.

Perform the numerical applications for the following machines:

– for a Vestas V80 wind turbine (Figure 3.2) for which $D = 80$ m, when the wind speed is 30 km/h. What is the energy theoretically produced in one year if we accept a load factor $f = 25\%$?

– for an offshore wind turbine with a diameter $D = 170$ m, for which the wind speed is 10 m/s and the load factor $f = 35\%$, carry out the same numerical applications;

– for an Openhydro marine current turbine (Figure 3.2) with a diameter $D = 16$ m in a tidal current with a speed of 2 m/s;

– How many wind turbines must a “wind farm” have to provide the capacity of a nuclear power plant unit?

Density of air is $\rho = 1.25$ kg/m^3, and density of water is $\rho = 1{,}000$ kg/m^3.

3.2.1. *Solution*

1) In the Betz model, the fluid particles only give up head when they pass through the wind turbine rotor since any parasitic head loss is neglected. The head therefore has the same value at any point of the surfaces S_1, S_{ext} and $S_{L,far}$ and on these surfaces the speed is the wind speed U_{wind}.

The pressure is hydrostatic since the flow is parallel on these surfaces away from the wind turbine (section 1.3). The same is true on the surface S_2. The piezometric pressure is what is called atmospheric pressure: $P_{atm} = p + \rho g z$.

Inside the flow tube surrounded by the surface $S_{L,near}$, the head and the velocity are also uniform in any discharge section perpendicular to the axis Ox. The piezometric pressure $p + \rho g z$ will therefore also be uniform, but its value may differ from the atmospheric pressure P_{atm} defined previously, as will be seen later, depending on the position x of this flow section.

2) A fluid particle passing between the blades transfers part of its head to the rotor of the wind turbine. The head in the section S_1 must therefore be greater than the head in the section S_2:

$$P_{atm} + \rho \frac{U_{wind}^2}{2} > P_{atm} + \frac{U_{wake}^2}{2}$$

The velocity U_{wake} in the section S_2 is therefore less than U_{wind}.

The conservation of the flow rate between the various sections is written as:

$$U_{wind} S_1 = US = U_{wake} S_2$$

Since $U_{wake} < U_{wind}$, then $S_2 > S_1$. There must therefore occur a widening of the flow tube passing through the surface of the wind turbine.

3) Since the only head loss undergone by the fluid particles is that transferred to the rotor of the wind turbine, Bernoulli's theorem can be applied between the section S_1 and the near upstream section. Hence:

$$P_{upstream} = P_{atm} + \rho\{U_{wind}^2 - U^2\}/2$$

By doing the same between the near downstream section and the section S_2:

$$P_{downstream} = P_{atm} + \rho\{U_{wake}^2 - U^2\}/2$$

As a result:

$$\Delta H = P_{upstream} - P_{downstream} = \rho\frac{U_{wind}^2 - U_{wake}^2}{2} = \rho\frac{(U_{wind} - U_{wake})(U_{wind} + U_{wake})}{2}$$

The rotor and alternator convert fluid particle pressure energy into electrical energy. The pressure increases between far upstream and near upstream. It decreases while passing through the blades, and then increases again between near downstream and far downstream.

The head, speed and pressure variations along the flow tube passing through the wind turbine are shown schematically in Figure 3.6. The nature of the conversion processes that take place successively, upstream of the wind turbine, when crossing the rotor, then downstream of the wind turbine, is explained in bold print at the bottom of the figure.

4) The power recovered by the wind turbine is written as:

$$\wp = \Delta H.Q = \rho\frac{U_{wind}^2 - U_{wake}^2}{2}US$$

5) The momentum theorem (section 1.8) allows us to calculate the force applied to the rotor in the direction Ox:

$$-\overrightarrow{F_{rotor}} = \iint_{S_d}(p + \rho gz)\vec{n}ds + \iint_{S_d}\rho(\vec{u}.\vec{n})\vec{u}ds$$

The surface surrounding the domain is here $S_d = S_1 + S_{ext} + S_{L,far} + S_2$.

Since $p + \rho g z = P_{atm}$ is constant over the closed surface S_d, the pressure integral is zero by virtue of Ostrogradski's theorem (see section 1.8). Only the integrals of the flow rate of momentum are to be calculated.

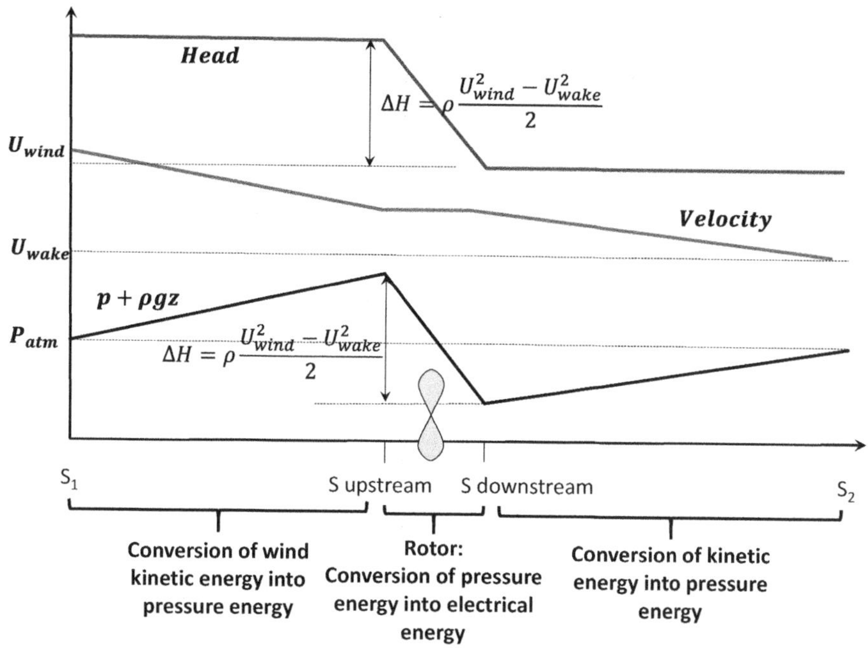

Figure 3.6. *Variation of head, flow rate velocity and piezometric pressure along a streamline crossing a wind turbine*

By noting $\vec{\imath}$ the unit vector according to the direction Ox, and since $\vec{n}$ is the normal outside the domain:

$$\iint_{S1} \rho(\vec{u}.\vec{n})\,\vec{u}ds = -\iint_{S1} \rho(U_{wind}\vec{\imath}.\vec{\imath})\,U_{wind}\vec{\imath}ds = -\rho U_{wind}^2 S_1\vec{\imath}$$

$$\iint_{S2} \rho(\vec{u}.\vec{n})\,\vec{u}ds = \iint_{S2} \rho(U_{wake}\vec{\imath}.\vec{\imath})\,U_{wake}\vec{\imath}ds = \rho U_{wake}^2 S_2\vec{\imath}$$

The integral on the surface $S_{L,far}$ is zero since $(\vec{u}.\vec{n}) = 0$ on this surface which envelops a flow tube.

For the integrals on the surface S_{ext}, we distinguish the upstream surface from the downstream surface, since the normal vector changes sign:

$$\iint_{Sext\ upstream} \rho(\vec{u}.\vec{n})\,\vec{u}ds = -\iint_{Sext\ upstream} \rho(U_{wind}\vec{\iota}.\vec{\iota})\,U_{wind}\vec{\iota}ds$$
$$= -\rho U_{wind}^2 S_{ext\ upstream}\vec{\iota}$$

$$\iint_{Sext\ downstream} \rho(\vec{u}.\vec{n})\,\vec{u}ds = \iint_{Sext\ downstream} \rho(U_{wind}\vec{\iota}.\vec{\iota})\,U_{wind}\vec{\iota}ds$$
$$= \rho U_{wind}^2 S_{ext\ downstream}\vec{\iota}$$

Flow rate conservation:

$$U_{wind}S_{ext\ upstream} = U_{wind}S_{ext\ downstream}$$

causes the two preceding integrals to cancel each other out.

The end result is therefore:

$$\overrightarrow{F_{rotor}} = \rho U_{wind}^2 S_1\vec{\iota} - \rho U_{wake}^2 S_2\vec{\iota} = \rho US\{U_{wind} - U_{wake}\}\vec{\iota}$$

We can legitimately wonder about the reasons which make us choose such a vast domain to apply the momentum theorem because the integrals on the surfaces $S_{ext} + S_{L,far}$ are zero. Many proofs of Betz's formula apply moreover the momentum theorem on the domain surrounded by the surface $S = S_1 + S_{ext} + S_{L,near}$. These calculations give the same result assuming that the pressure $p + \rho gz = P_{atm}$ on the flow tube surface $S_{L,near}$. But this is contradictory with the conclusions of question 3 (Figure 3.6).

6) Applying the momentum theorem to the rectangular volume surrounding the wind turbine also determines the force applied to the rotor:

$$-\overrightarrow{F_{rotor}} = -\iint_{Supstream} \{\rho(U\vec{\iota}.\vec{\iota})U\vec{\iota} + P_{upstream}\vec{\iota}\}ds$$
$$+ \iint_{Sdownstream} \{\rho(U\vec{\iota}.\vec{\iota})U\vec{\iota} + P_{downstream}\vec{\iota}\}ds$$

The integral over the side surface of the cylinder is zero because there is zero flow rate through that surface and the pressure force is perpendicular to $\vec{\iota}$. Hence the result:

$$\overrightarrow{F_{rotor}} = \{P_{upstream} - P_{downstream}\}S\vec{\imath} = \left\{\rho\frac{U_{wind}^2}{2} - \rho\frac{U_{wake}^2}{2}\right\}S\vec{\imath}$$
$$= \rho\frac{(U_{wind} - U_{wake})(U_{wind} + U_{wake})}{2}S\vec{\imath}$$

The equality of the expressions of $\overrightarrow{F_{rotor}}$ obtained in questions 5 and 6 leads to:

$$U = \frac{U_{wind} + U_{wake}}{2}$$

By introducing this quantity into the expression of the power established in question 4, the power only depends on a single unknown $a = U_{wake}/U_{wind}$ according to a relation of the form:

$$\wp(a) = \rho\frac{U_{wind}^3}{4}S(1 + a)^2(1 - a)$$

7) The power recovered is maximum when:

$$\frac{d\wp}{da} = \rho\frac{U_{wind}^3}{4}S\frac{d\{(1 + a)^2(1 - a)\}}{da} = 0$$

which is obtained for $a = 1/3$.

Betz's formula gives the maximum power $\wp(a = 1/3)$ that can be recovered by a wind turbine:

$$\wp_{max} = \rho\frac{U_{wind}^3}{2}S\frac{16}{27}$$

The power coefficient:

$$c_p = \frac{\wp_{max}}{\rho\frac{U_{wind}^3}{2}S} = \frac{16}{27}$$

is the maximum efficiency of a wind turbine. The definition of the power coefficient in this form is naturally understood if one notes that $\rho U_{wind}^2/2$ is the maximum energy per unit volume that can be recovered by a wind turbine and that $U_{wind}S$ is the maximum flow rate that can pass through the wind turbine.

The parameter $a = U_{wake}/U_{wind}$ can vary between 0 and 1.

If $a = 1$, the fluid particles do not transfer any energy to the rotor and the flow tube is a circular cylindrical tube with a flowing surface S.

If $a = 0$, the kinetic energy of the wind is fully recovered by the rotor but the flow rate is reduced so that the power recovered is less than in the case where $a = 1/3$.

The Betz model clarifies the role of flow widening and the importance of the wake in the energy recovery process. But it does not model the energy conversion itself, since the dynamics of the rotor (its blades and its rotation) is not taken into account in the model. A rotor model is proposed in the exercise of section 3.3.

Betz's formula allows a simple and quick estimation of the power of a wind turbine. It is often used because it gives an order of magnitude close to reality.

Betz's formula also applies to a marine turbine because the flows are similar. Since the density of water is 800 times greater than that of air, marine current turbines are smaller in size and more massive than wind turbines because they have to withstand greater mechanical stress. Despite their smaller size and weaker sea currents (at most 2–3 m/s), they can harvest significant power thanks to the density of the water.

The area swept by the blades of the wind turbine is related to the diameter of a wind turbine by the relationship $S = \pi D^2/4$.

A wind turbine manufacturer usually specifies a range of wind speeds in which the wind turbine operates and a nominal power. With Betz's formula, this nominal power can be associated with a wind speed. The load factor f for a given site is the ratio of the annual energy (in Wh) recovered to the nominal power (in W) of the wind turbine multiplied by the number of hours in a year (8,760), hence $E_{annual} = 8{,}760 \times f \times \wp_{nom}$.

The numerical applications give the following results (Table 3.2).

A 1,000 MW nuclear power plant reactor is equivalent to approximately 1,000 VESTAS 80 wind turbines in terms of power, but 250 wind turbines in terms of energy recovered, due to the intermittency of the winds evaluated by the load factor.

	Reference	D (m)	U_{wind} (m/s)	$\wp_{max}$ (MW)	f (–)	E_{annual} (GWh)
Onshore wind turbine	VESTAS 80	80	8.33	1.08	0.25	2.36
Offshore wind turbine	–	170	10	8.41	0.35	25.77
Marine current turbine	Openhydro	16	2	0.48	0.5	2.09

Table 3.2. *Numerical applications of Betz's formula to estimate the power and annual energy that can be recovered by two wind turbines and one marine current turbine*

The table quantifies the difference between an onshore wind turbine and an offshore wind turbine. Offshore wind turbines reach dimensions of the order of twice those of an onshore wind turbine. Winds are generally stronger at sea than on land. As the power increases with the cube of the wind speed, the effect is significant. Finally, load factors are generally higher at sea than on land.

The Openhydro marine current turbine (EDF-DNS) shown in Figure 3.2 has a nominal power of 500 kW for a diameter of 16 m. For a current of 2 m/s, the Betz formula evaluates the power recovered at 480 kW.

3.3. Exercise: wind turbine rotor model

This model sizes the rotor of a wind turbine for which the angle of the blades is chosen. The speed of rotation is determined so as to maximize the energy recovered.

The configuration (Figure 3.7) is modified compared to the exercise dedicated to Betz's formula (Figure 3.5) by taking into account the rotation of the flow generated downstream of the rotor. The speed component u_θ, in the cylindrical coordinate frame whose axis Oz coincides with the axis of rotation of the rotor, is worth $u_\theta = 0$ upstream of the wind turbine (sections S_1 and S just in front of the turbine), while it is non-zero in the section S just behind the rotor and in section S_2. Due to a non-zero rotational component, the piezometric pressure in the section S at the rear of the rotor varies with the radial distance r from the Oz axis (i.e. $p_{downst}(r)$).

In the green domain, between the upstream and downstream sections S, the flow is assumed to be cylindrical: the radial velocity component is $u_r = 0$. It is further assumed that the axial velocity u_z component is uniform in the four sections S_1, S upstream and downstream, as well as in the section S_2. Note the wind speed $u_z = U_{wind}$ in the section S_1 while $u_z = U$ in the upstream section S.

The right part of Figure 3.8 shows a plan view of the rotor. In the left part of this figure, we schematize the flow around a rotor blade on the surface of the circular cylinder of radius r. The blade segment is a flat plate (dark bold segment) that makes an angle $\alpha(r)$ with the Oz direction of the wind. This representation makes it possible to represent the velocity components according to the unit vectors $\vec{e}_\theta$ and $\vec{e}_z$ in three planes, that is, 1, 2 and 3. We have already plotted the velocity in the fixed frame in section 1 (surface S_1) and indicated its velocity components on the figure. It will be completed for plans 2 and 3 by answering the first question. The speed of the blade element $\vec{u}_p(r) = \omega r \vec{e}_\theta$ is also represented.

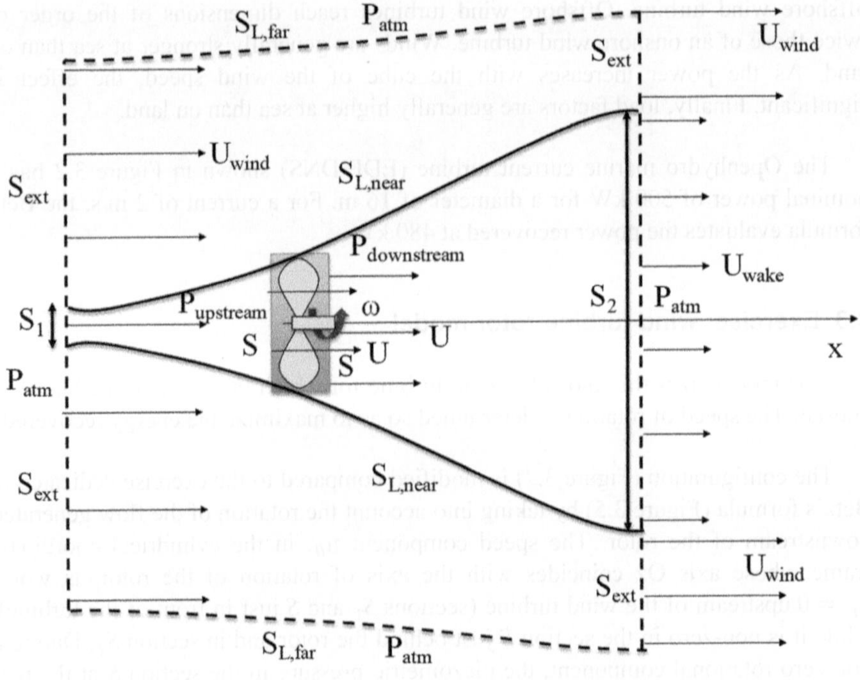

Figure 3.7. *Representation of the flow around a wind turbine. The rotation of the rotor is taken into account. For a color version of this figure, see www.iste.co.uk/mory/mechanical.zip*

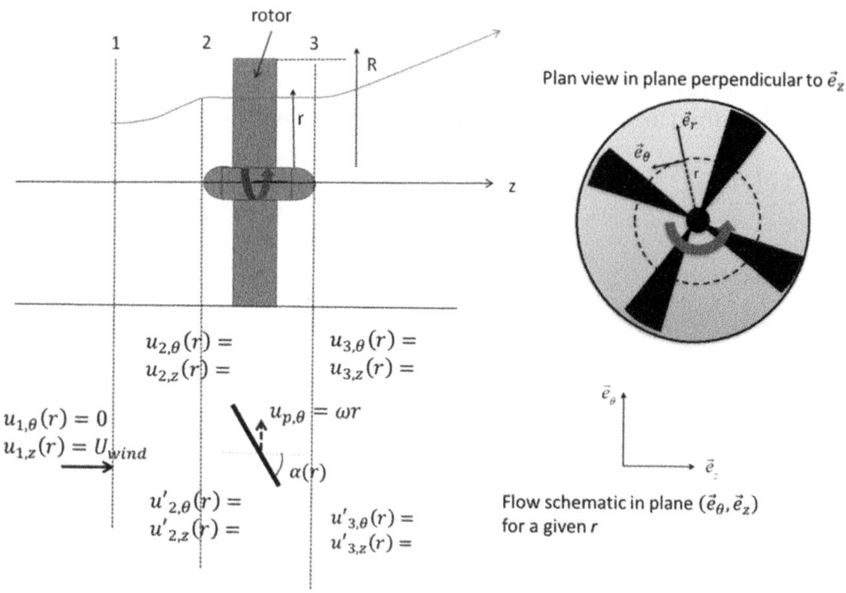

Figure 3.8. *Plan view of the wind turbine in the plane perpendicular to the wind direction, and velocity triangle in the plane* $(\vec{e}_\theta, \vec{e}_z)$

1) In this question, we determine the velocity triangle for the rotor element contained in the annular cylinder of radius r and thickness dr. To do this, indicate in Figure 3.8 the speed components in planes 2 and 3 in the fixed frame (speed $\vec{u}$) and in the frame moving with the blade (speed $\overrightarrow{u'}$) and plot the speed vectors in the plan $(\vec{e}_\theta, \vec{e}_z)$. These velocity components must verify the conservation of flow rate and the Joukovski condition.

2) In an inviscid fluid, an irrotational flow must remain so. In a cylindrical coordinate system (Figure 1.3), the curl of velocity is written as:

$$\overrightarrow{Rot}\vec{u} = \begin{pmatrix} \frac{1}{r}\frac{\partial u_z}{\partial \theta} - \frac{\partial u_\theta}{\partial z} \\ \frac{\partial u_r}{\partial z} - \frac{\partial u_z}{\partial r} \\ \frac{1}{r}\frac{\partial (r u_\theta)}{\partial r} - \frac{1}{r}\frac{\partial u_r}{\partial \theta} \end{pmatrix}$$

We verify that the flow is irrotational in plane 1 where the wind is uniform. If the rotor works well (in particular without producing a head loss added to the head transmitted to the rotor), the flow in the fixed frame must also be irrotational in plane 3. We deduce from this, with the kinematic properties previously established, that:

$$\frac{1}{r}\frac{\partial(ru_{3,\theta})}{\partial r} = 0$$

As a result:

$$u_{3,\theta}(r) = -\frac{\Gamma}{2\pi r}$$

Establish that the variation in the angle of the blades must be such that:

$$tg\{\alpha(r)\} = a'\frac{R}{r} + b'\frac{r}{R}$$

to verify the condition of Joukovski and the irrotationality of the flow in the fixed frame.

3) The parameters a' and b' are dimensionless. Deduce from the above two equations linking a', b', U, Γ and the radius of the blades R. Apply the momentum theorem in the annular cylinder of radius r and thickness dr that contains the rotor element between planes 2 and 3. Why should it be applied using the velocity components in the moving frame? Then deduce the power recovered by the rotor element contained in this volume.

4) The power recovered by the rotor is transferred by the fluid particles whose head (estimated using the velocity components in the fixed frame) decreases between planes 2 and 3. Use this property to express the radial variation of pressure $p_{downstream}(r)$ in plane 3. Applying the boundary condition $p_{downstream}$ $(r = R) = P_{atm}$ at the tip of the blade, deduce a relationship between the speeds U and U_{wind}. We will use the relations established in question 2 to write this relation as a function of a' and b'.

5) Calculate the total power recovered by the rotor as a function of U_{wind}, R, a' and b'.

6) Determine the values of a' and b' which maximize the power recovered by the rotor and determine the maximum power that can be recovered. What is the angular rotation speed of the rotor?

Keeping $a'b' = 1$, express the variations of the power coefficient c_p and of the tip speed ratio λ as a function of b'. Superimpose the relationship $c_p(\lambda)$ on Figure 3.3 and comment on the diagram.

7) For three values $b' = 1$, $b' = 13$ and $b' = 30$, keeping $a'b' = 1$, determine the variation of the blade angle $\alpha(r/R)$ as well as the variation of the angle of incidence $\beta(r/R)$ made by the speed (in the moving frame) with the blade in plane 2. Plot the two curves for the three conditions and comment.

3.3.1. *Solution*

1) As in the Betz model, the speed U in plane 2 immediately upstream of the rotor is lower than the wind speed in plane 1, due to the widening of the flow.

By conservation of the flow rate, since the flow is cylindrical between planes 2 and 3, $u_{3,z} = u_{2,z} = U$.

The flow has no rotation component upstream of the rotor: $u_{2,\theta} = 0$.

The Joukovski condition determines the velocity in the moving frame in plan 3:

$$\overrightarrow{u'}_3(r) = -Utg\alpha\vec{e}_\theta + U\vec{e}_z$$

The speeds in the fixed reference frame and in the mobile reference frame are linked by the composition of speeds: $\vec{u} = \omega r\vec{e}_\theta + \overrightarrow{u'}$.

We reproduce Figure 3.8 on Figure 3.9 after having completed it.

2) The angle $\alpha(r)$ is determined by equalizing the irrotational form of $u_{3,\theta}$ with its form resulting from the Joukovski condition (question 1):

$$u_{3,\theta}(r) = -\frac{\Gamma}{2\pi r} = -Utg\{\alpha(r)\} + \omega r$$

The angle between the blade and the direction of the wind therefore varies according to:

$$tg\{\alpha(r)\} = \frac{\Gamma}{2\pi U R}\frac{R}{r} + \frac{\omega R}{U}\frac{r}{R} = a'\frac{R}{r} + b'\frac{r}{R}$$

where:

$$b' = \frac{\omega R}{U} \quad \text{and} \quad a' = \frac{\Gamma}{2\pi UR}$$

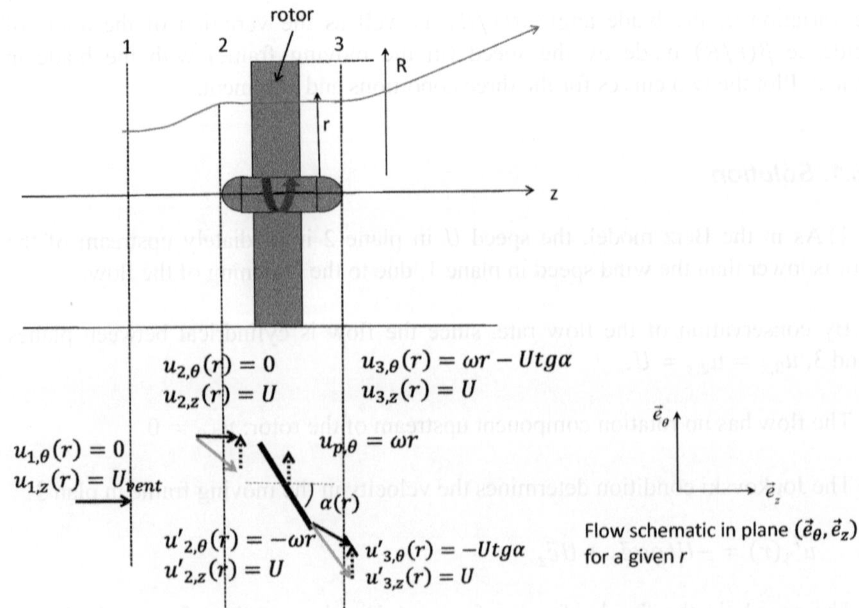

Figure 3.9. *Velocity triangle in the plane $(\vec{e}_\theta, \vec{e}_z)$ (speed of the blade in dashed line, speed in the fixed frame in black, speed in the moving frame in gray)*

The geometry of the rotor is determined by the choice of the values a' and b'. The two previous relationships link three unknown parameters at this stage: the angular speed of rotation ω, Γ (called circulation) and the flow rate velocity through the wind turbine U.

3) The momentum theorem (section 1.8) must use the velocities in the rotating frame, in which the flow is stationary, in the annular volume of radius r and thickness dr between plans 2 and 3. The force applied on the element of rotor in the volume is deduced:

$$\overrightarrow{dF} = -\{\rho((-\omega r\vec{e}_\theta + U\vec{e}_z).(-\vec{e}_z))(-\omega r\vec{e}_\theta + U\vec{e}_z) + P_2(-\vec{e}_z)\}2\pi r dr$$
$$- \{\rho((-Utg\alpha\vec{e}_\theta + U\vec{e}_z).(\vec{e}_z)) + P_3(\vec{e}_z)\}2\pi r dr$$

Hence:

$$\overrightarrow{dF} = dF_\theta \vec{e}_\theta + dF_z \vec{e}_z = \rho U\{-\omega r + Utg\alpha\}2\pi r dr \vec{e}_\theta + \{P_2 - P_3\}2\pi r dr \vec{e}_z$$

The power recovered by the rotor element contained in the annular cylinder of radius r and thickness dr is therefore:

$$d\wp = \{dF_\theta \vec{e}_\theta + dF_z \vec{e}_z\}.\omega r \vec{e}_\theta = \omega r dF_\theta = \omega r \rho U\{-\omega r + Utg\alpha\}2\pi r dr$$

Since (question 2):

$$u_{3,\theta}(r) = -Utg\{\alpha(r)\} + \omega r = -\frac{\Gamma}{2\pi r}$$

$$d\wp = \omega\rho\frac{\Gamma}{2\pi}(U2\pi r dr) = \omega\rho\frac{\Gamma}{2\pi}dQ$$

The total power recovered by the rotor is then deduced by integration on the flowing section through the rotor:

$$\wp = \omega\rho\frac{\Gamma}{2\pi}Q$$

4) Since:

$$d\wp = \omega\rho\frac{\Gamma}{2\pi}dQ = \Delta H dQ$$

we deduce that:

$$\Delta H = \omega\rho\frac{\Gamma}{2\pi}$$

A remarkable and important property is that ΔH is identical for all particles passing through the rotor (regardless of radial position r). It makes relevant the assumption made in the statement that the delivery speed U is uniform in the rotor. If the head loss were greater for certain radial positions r, the flow rate passing through these volumes would be reduced to pass more flow through other sections of the rotor where the head loss is lower. The speed would therefore no longer be uniform in the delivery section S.

The head variation between planes 2 and 3 is also expressed as a function of the pressures and speeds (in the fixed frame):

$$\Delta H = \left\{P_2 + \rho\frac{U^2}{2}\right\} - \left\{P_3(r) + \rho\frac{(\omega r - Utg\alpha)^2}{2} + \rho\frac{U^2}{2}\right\} = \omega\rho\frac{\Gamma}{2\pi}$$

As the head is conserved between planes 1 and 2:

$$P_2 + \rho\frac{U^2}{2} = P_{atm} + \rho\frac{U_{wind}^2}{2}$$

It is then simpler to replace the head in plane 2 by its expression in plane 1. Hence:

$$P_{atm} + \rho\frac{U_{wind}^2}{2} - P_3(r) - \rho\frac{(\omega r - Utg\alpha)^2}{2} - \rho\frac{U^2}{2} = \omega\rho\frac{\Gamma}{2\pi}$$

then:

$$P_3(r) - P_{atm} = \rho\frac{U_{wind}^2}{2} - \rho\frac{(\omega r - Utg\alpha)^2}{2} - \rho\frac{U^2}{2} - \omega\rho\frac{\Gamma}{2\pi}$$

With the boundary condition that the pressure at the blade tip is the atmospheric pressure ($P_3(r = R) = P_{atm}$):

$$0 = \rho\frac{U_{wind}^2}{2} - \rho\frac{(\omega R - Utg\alpha)^2}{2} - \rho\frac{U^2}{2} - \omega\rho\frac{\Gamma}{2\pi}$$

With the results of question 1:

$$u_{3,\theta}(R) = -\frac{\Gamma}{2\pi R} = -Utg\{\alpha(R)\} + \omega R$$

we deduce:

$$0 = \rho\frac{U_{wind}^2}{2} - \rho\frac{\Gamma^2}{2(2\pi R)^2} - \rho\frac{U^2}{2} - \omega\rho\frac{\Gamma}{2\pi}$$

Finally, with the results of question 2:

$$b' = \frac{\omega R}{U} \quad \text{and} \quad a' = \frac{\Gamma}{2\pi U R}$$

we deduce:

$$0 = \frac{U_{wind}^2}{2} - a'^2\frac{U^2}{2} - \frac{U^2}{2} - a'b'U^2$$

We get ultimately:

$$U = \frac{U_{wind}}{\sqrt{1 + a'^2 + 2\,a'b'}}$$

The values a' and b' thus determine the relation linking U and U_{wind}.

5) Taking the expression of the total power obtained in question 3, using the expressions of question 1 and the previous relation, the total power recovered by the rotor is expressed as a function of the speed U_{wind}, the radius of the rotor R and blade angle parameters a' and b':

$$\wp = \rho U_{wind}^3 \frac{a'b'}{(1 + a'^2 + 2\,a'b')^{3/2}}\pi R^2$$

6) The power recovered is maximum when:

$$\frac{d\wp}{d(a'b')} = 0 \quad \text{and} \quad \frac{d\wp}{da'} = 0$$

These conditions are obtained when $a' = 0$ and $a'b' = 1$ such that $b' \to \infty$.

For these conditions (except in $r = 0$ which is a singular point), the blades are perpendicular to the wind direction (angle α is 90 °) and $\omega \to \infty$.

The maximum power that can be recovered by the wind turbine is then:

$$\wp_{max} = \rho U_{wind}^3 \frac{1}{3^{3/2}}\pi R^2$$

Keeping $a'b' = 1$, the variation of the blade angle is written as:

$$tg\{\alpha(r)\} = \frac{1}{b'}\frac{R}{r} + b'\frac{r}{R}$$

The power is:

$$\wp(b') = \rho U_{wind}^3 \frac{b'^3}{(1+3b'^2)^{3/2}} \pi R^2$$

and the power coefficient:

$$c_p(b') = \frac{\wp(b')}{\rho \frac{U_{wind}^3}{2} \pi R^2} = \frac{2b'^3}{(1+3b'^2)^{3/2}}$$

As $b' = \omega R/U$, the tip speed ratio is written as:

$$\lambda = \frac{\omega R}{U_{wind}} = b'\frac{U}{U_{wind}} = \frac{b'}{\sqrt{1+a'^2+2\,a'b'}} = \frac{b'^2}{\sqrt{1+3b'^2}}$$

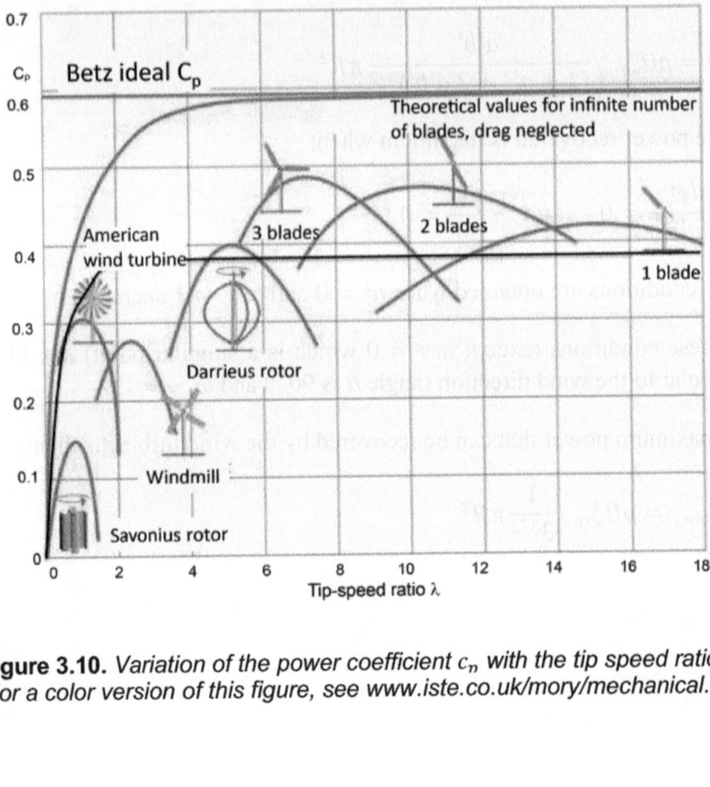

Figure 3.10. *Variation of the power coefficient c_n with the tip speed ratio λ. For a color version of this figure, see www.iste.co.uk/mory/mechanical.zip*

The variations $c_p(\lambda)$ are superimposed as a black curve in Figure 3.10 on the graph shown in Figure 3.3. The graph leads to the following comments:

– In this exercise, the wind turbine is sized by fixing the blade angles with the choice of parameters a' and b'. We deduce the output speed U in the plane of the wind turbine, the angular speed of rotation ω and the power recovered $\wp$.

– The curves in brown, for the different wind turbines, make it possible to determine optimal operating conditions, those for which the power coefficient (and therefore the recovered power) is maximum. The optimal rotor speed is deduced from the corresponding value of λ.

– The optimization performed in this exercise is different from that associated with the brown curves shown in the figure. For the latter, the blade angles are fixed and the variation of the power coefficient is plotted when the speed of rotation varies. In our model, the variation of a' and b' induces a modification of the speed of rotation ω so as to preserve the irrotationality of the flow.

– The wind turbine designed in this exercise is similar to an “American” wind turbine (infinite number of blades). The best operating condition is theoretically an infinite speed of rotation of the rotor. This operating condition is not recommended in practice because the angle of incidence becomes very small (see question 7) and this operating condition is not stable. A slight change in the wind speed is enough to reverse the direction of rotation of the rotor. The curve shows that the power coefficient increases slightly beyond $\lambda > 2$. American wind turbines are generally used with a value of λ around 1.

– The maximum power coefficient ($c_p = 0.39$) for the wind turbine dimensioned in this exercise is lower than the limit value of the Betz formula, as expected. Wind turbines with three blades have a best efficiency $c_p = 0.48$ for $\lambda = 7$.

7) Keeping $a'b' = 1$, the variation of the blade angle is written as:

$$tg\{\alpha(r)\} = \frac{1}{b'}\frac{R}{r} + b'\frac{r}{R}$$

With $x = r/R$:

$$\alpha(x) = Arctg\left(\frac{1 + b'^2x^2}{b'x}\right)$$

$$\cos\alpha = \sqrt{\frac{1}{1 + tg^2\alpha}} = \sqrt{\frac{b'^2x^2}{1 + 3b'^2x^2 + b'^4x^4}}$$

The unit vector tangent to the blade is $\vec{t}_{blade} = -\sin\alpha\vec{e}_\theta + \cos\alpha\vec{e}_z$.

The unit vector parallel to the relative velocity of the flow with respect to the blade, in plane 2, is:

$$\vec{t'}_{vel} = \frac{-\omega r}{\sqrt{U^2+\omega^2 r^2}}\vec{e}_\theta + \frac{U}{\sqrt{U^2+\omega^2 r^2}}\vec{e}_z$$

The angle of incidence β of the flow with respect to the blade is therefore:

$$\cos\beta = \vec{t}_{blade}.\vec{t'}_{vel} = \{-\sin\alpha\vec{e}_\theta + \cos\alpha\vec{e}_z\}.\left\{\frac{-\omega r}{\sqrt{U^2+\omega^2 r^2}}\vec{e}_\theta + \frac{U}{\sqrt{U^2+\omega^2 r^2}}\vec{e}_z\right\}$$

$$\cos\beta(x) = \cos\alpha\frac{\omega r tg\alpha + U}{\sqrt{U^2+\omega^2 r^2}} = \cos\alpha\frac{b'x.tg\alpha + 1}{\sqrt{1+b'^2x^2}}$$

with $b' = \omega R/U$.

Hence:

$$\cos\beta(x) = \frac{b'x'}{\sqrt{1+3b'^2x^2+b'^4x^4}}\frac{2+b'^2x^2}{\sqrt{1+b'^2x^2}}$$

The plots of angles $\alpha(r)$ and $\beta(r)$ are shown in Figure 3.11 for three values of b':

– For $b' = 1$, the angle of incidence $\beta(r)$ of the flow with respect to the blade is greater than 15° all along the blade. The stall is predictable.

– For $b' = 13$, $\beta(x) < 15°$ for $x > 0.1$. Increasing b' reduces the angle of incidence. The flow in the vicinity of the hub is singular. The angle of incidence is large so that flow separation on the blade extrados is predictable near the hub.

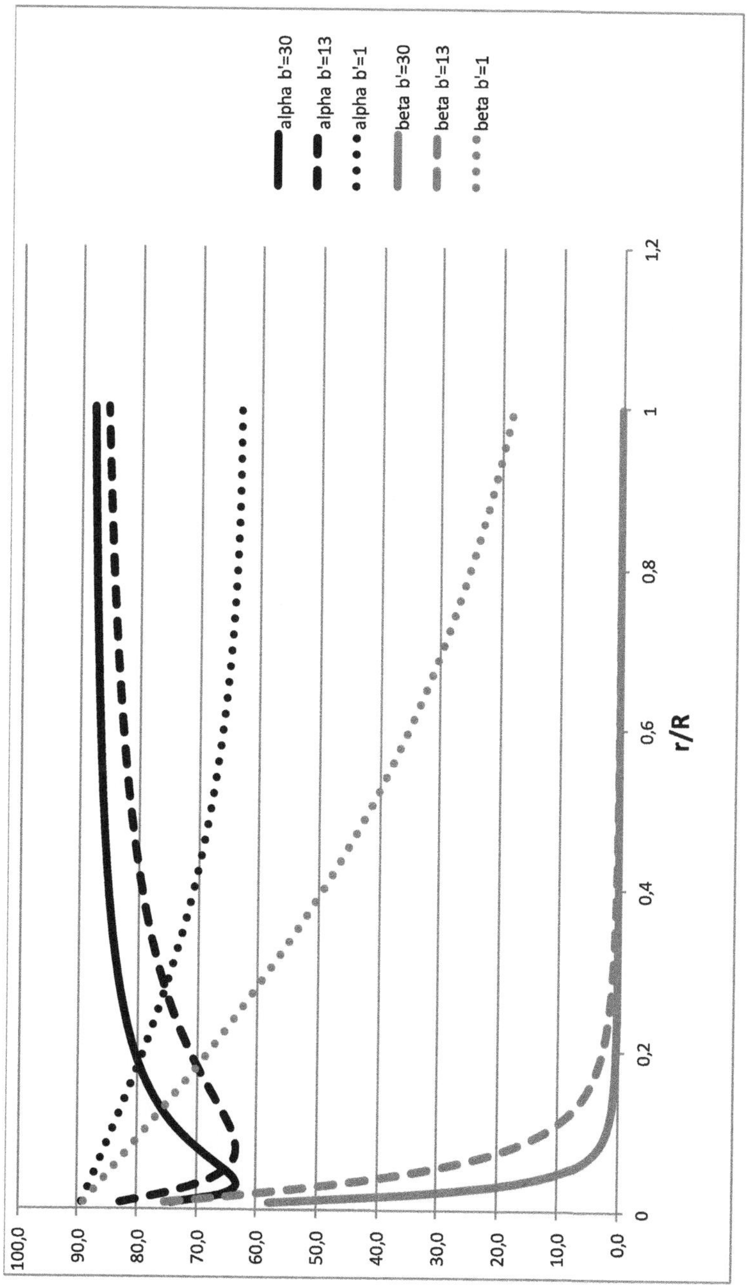

Figure 3.11. *Radial variation of the angle $\alpha(r)$ of the blade with the direction of the wind (solid lines) and of the relative angle of incidence $\beta(r)$ (dotted lines) for three values of b' and a' b' = 1*

4

Tidal Energy

4.1. General information and status of the sector

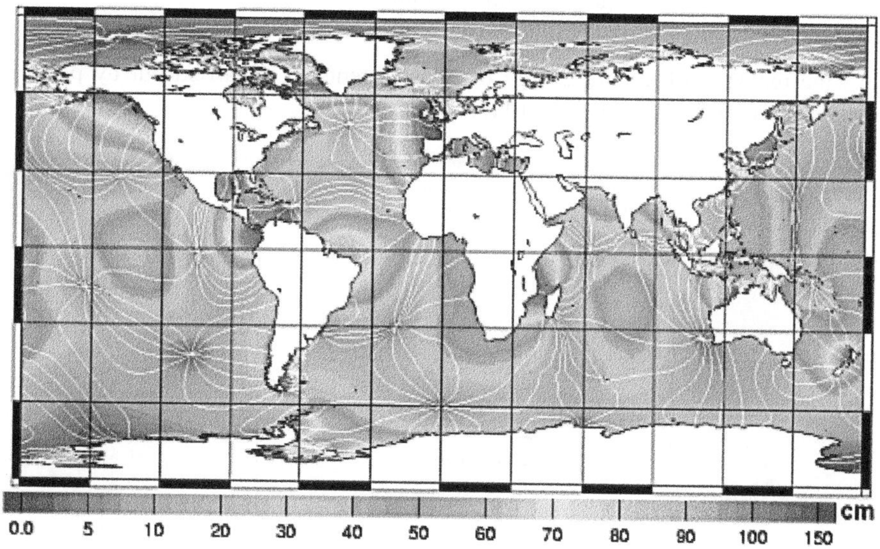

Figure 4.1. *Distribution of tidal amplitudes (in cm) at the surface of seas and oceans*[1]*.*
For a color version of this figure, see www.iste.co.uk/mory/mechanical.zip

Tides are the displacements of water masses generated by variations in the forces of gravity applied on the Earth by the Moon in rotation around the Earth. They

1 Reproduced with the permission of F. Lyard and D. Allain, SIROCCO/INSU, Laboratory of Space Geophysical and Oceanographic Studies, France.

produce vertical displacements of sea level, associated with variations in potential energy, and tidal currents, associated with kinetic energy. Tides are only significant in an ocean-sized basin, such as the Atlantic Ocean or the Pacific Ocean. In a lake or a sea, even as large as the Mediterranean Sea, the tides are of low amplitude.

Tidal energy is theoretically considerable. D. Aelbrecht[2], using data from the World Energy Council (2003), assesses its power at 3,000 GW and the annual energy available at 25,000 TWh. As shown in Figure 4.1, tidal ranges are only significant on some continental shelves, in particular on the continental shelf of the Atlantic Ocean bordering the western coasts of Europe, in the English Channel, around Ireland and the British Isles.

The recovery of tidal energy is considered in two ways:

– the recovery of potential energy related to variations in sea level. The system consists of building a dam to close an estuary in which the tidal range is high. The flow of tidal water through the dam is controlled by turbines, which recover a fraction of the energy, at rising tide and at falling tide;

– the recovery of kinetic energy of tidal currents. It is carried out by placing marine current turbines in the water layer.

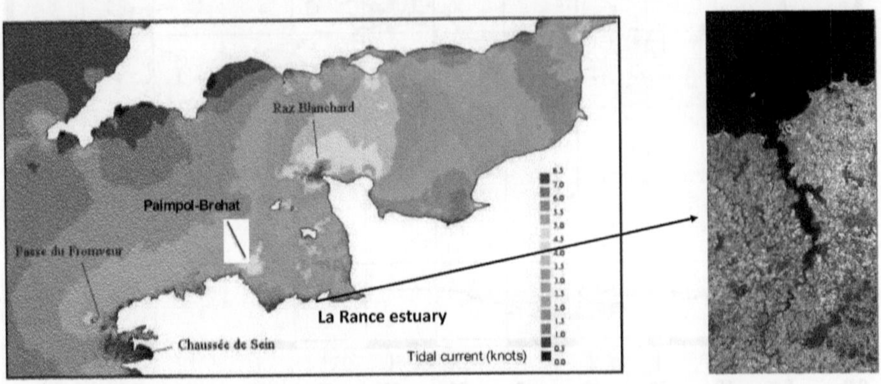

Figure 4.2. *Map of the amplitude of tidal currents (in knots) in the English Channel, calculated by EDF R&D using Telemac 2D software[3]. For a color version of this figure, see www.iste.co.uk/mory/mechanical.zip*

The kinetic energy of tidal currents is significant only on continental shelves above which the water layer depth is shallow. The places of interest for exploiting

2 See Aelbrecht (2015).

3 Source: © EDF/D. Aelbrecht. All rights reserved.

tidal energy are therefore few in number. On the map of tidal currents shown in Figure 4.2, we locate various sites in the English Channel that are important for the study and exploitation of tidal energy:

– the Rance estuary, an aerial view of which is shown on the right;

– the Paimpol Bréhat site, which hosts a marine current turbines test site;

– the Alderney Race area, at the tip of the Cotentin and its strong tidal currents.

The operation of marine current turbines is similar to that of wind turbines. Exercises 3.2 (Betz formula) and 3.3 (wind turbine rotor) are relevant to evaluate the recovered power using the power coefficient:

$$c_p = \frac{\wp}{\frac{1}{2}\rho U_{current}^3 \pi R^2}$$

$\wp$ designates the power recovered by the marine current turbine, R is the radius of its rotor and $U_{current}$ is the speed of the current. The formula emphasizes that the energy resource of a marine current turbine is the kinetic energy of the sea current.

The differences between marine current turbines and wind turbines are identified by analyzing this formula. For marine current turbines, the density of water is about 800 times the density of the air in which a wind turbine operates. The effect of the density on the power recovered schematically compensates for the reduction of the speed, the speed of the sea currents (of the order of 1–2 m/s) being almost 10 times lower than the speed of the wind. Although the operating principles are similar, the appearance of a marine current turbine differs from that of a wind turbine. The surface of the blades of a marine current turbine is often greater than for a wind turbine (Figure 3.2). This makes it possible to reduce the stresses applied to the materials and thus to ensure the mechanical resistance of the rotor, because the force applied to the rotor of a marine current turbine is much greater than that applied to the rotor of a wind turbine. The rotor diameters of marine current turbines are significantly smaller than those of wind turbines currently being built, with regard to the thickness of the layer of water in which they are immersed and in order to ensure their mechanical strength.

The second difference between marine current turbines and wind turbines concerns the energy resource. For a wind turbine, the wind speed is a random variable over time of which only the statistical characteristics can be known. For a marine current turbine, the value of the tidal current at the point where the marine current turbine is installed can be determined with good precision. To a lesser extent, this remains true for a current turbine installed in a river.

Figure 4.3 shows a Hydroquest turbine, whose blades rotate around a vertical axis. An advantage of this technology, in the maritime context, is that the axis of the machine does not rotate to align with the sea current whose direction turns during a tide. The Hydroquest marine current turbine is the first at sea to have been connected to the electricity grid (June 2019) during its industrial test phase on the Paimpol-Bréhat site. The development of the marine current turbine sector has now reached, for the most advanced operators, the stage of industrial prototypes tested in operational conditions.

a) b)

Figure 4.3. *(a) Hydroquest vertical axis turbine (width 25 m, height 14 m, 1,400 tons, nominal power 1 MW); (b) its immersion on the Paimpol-Bréhat site in April 2019*[4]

Figure 4.4 shows an aerial view and diagram of the Rance plant dam. By controlling the flow passing through the turbines set up in the dam, the difference in level between the sea and the estuary is monitored. The turbines recover part of the difference in potential energy between the two sides of the dam. The exercise proposed in section 4.2 is a simplified model of the Rance tidal power plant. The optimization of the device, envisaged to maximize the energy recovered during a tidal cycle, is studied. A large photo gallery[5] allows viewing the Rance tidal power plant.

The sizing parameters of the Rance power plant are as follows:

– a 20 km long estuary, 22 km^2 in area forming a 184 hm^3 retention basin;

– tides with an average amplitude of 8 m. The maximum amplitude reaches 13.5 m.

4 Available at: www.hydroquest.fr.

5 Available at: http://www.hydrelect.info/articles.php?lng=fr&pg=295.

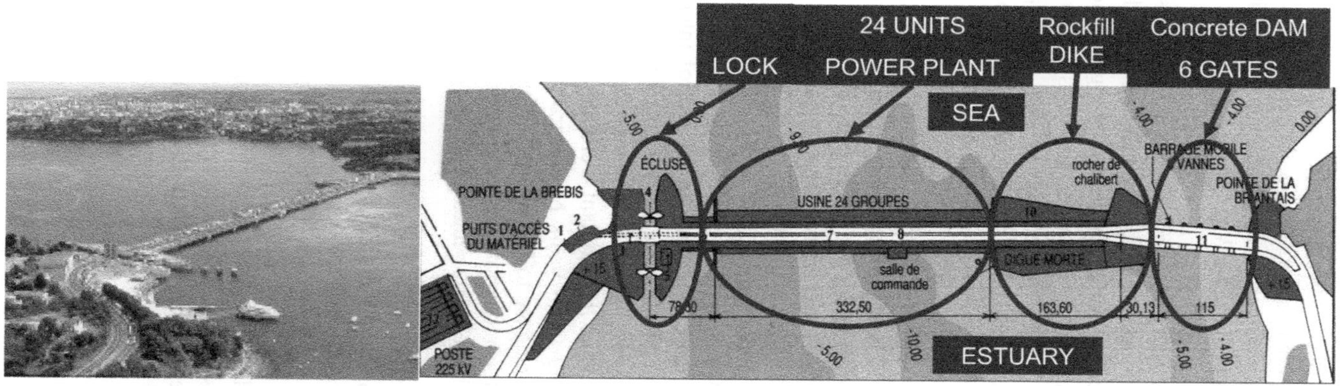

a) b)

Figure 4.4. *The Rance tidal power plant: (a) aerial view; (b) diagram (source: © EDF/D. Aelbrecht. All rights reserved)*

The tidal power plant is equipped with 24 reversible Kaplan turbine units, each with a nominal power of 10 MW. The power of the plant is 240 MW. Annual production is 538 GWh. The load factor is 25%. In other words, the installation should work a quarter of the time over a year if operated at maximum power. The cost of electricity production is €0.12 per kWh.

The Rance plant was inaugurated in 1966. The technology has proven itself for more than 55 years, showing its robustness for controlling the risks of corrosion in a saline environment in particular. Although the technology is industrially mature, one had nevertheless to wait until 2011 to see the construction of a facility of comparable size (252 MW) at Siwha in Korea. The other plants using this technology are located in Annapolis (United States, 1984, 20 MW), Xiangxa (China, 1980, 3.2 MW) and Kislaya Guba (Russia, 2004, 0.5 MW).

The number of estuaries exposed to strong tides is not enough to explain the limited number of plants of this type. Several projects on the Severn estuary in the United Kingdom have been considered, but never materialized. The environmental impacts are causing strong opposition to the construction of a tidal power plant closing off an estuary. For the Rance plant, the construction of the dam and the operation of the tidal power installation led to a 40% reduction in the amplitude of the tides inside the estuary basin[6]. The exercise proposed in section 4.2 explains why. The construction of the dam results in a reduction of intertidal zones (areas above water level at low tide and underwater at high tide) and a modification of wetlands with their consequences in terms of fauna, flora and biodiversity. The environmental consequences of the Rance plant were amplified by the fact that the estuary remained closed during the 3 years of construction (1963–1966). It took 10 years for a new ecological balance to be reached[7], which obviously differs from the previous balance due to the reduction in the amplitude of the tides inside the basin. The dam is permeable; biological species pass through turbines and sluices.

Another impact of the construction of the dam is the increase in sedimentation in the upper part of the estuary. EDF[8] recognizes the phenomenon, while arguing that it is also observed in neighboring estuaries free of comparable equipment. This sedimentation could therefore not be entirely due to the tidal power plant.

The present chapter includes a design exercise for a Kaplan turbine (section 4.3). The dimensioning parameters of this turbine correspond to the Kaplan turbines of the Rance plant. The exercise uses the tools introduced in Chapter 2. The approach implemented is similar to that followed in the exercise of section 3.3.

---

6 See Aelbrecht (2015).

7 Ibid.

8 Ibid.

Several operators are active in France[9] and the United Kingdom[10] to relaunch tidal power plant projects based on the concepts of *marine current turbines farm, tidal garden* or *tidal lagoon*. Figure 4.5 schematizes a *tidal lagoon* where marine current turbines are placed in channels surrounding a closed basin so as to recover the kinetic energy of the tidal currents. This configuration avoids obstructing an estuary by a dam and the marine current turbines are placed in a less confined space than are the turbines in a dam. However, when there are a large number of marine current turbines, the head loss of the fluid particles as they pass through the turbines leads to a reduction in the speed of the currents, through an effect that some call a mechanism to avoid passing through the turbine. The flow is diverted to adjacent spaces where water flows more freely. This phenomenon is modeled in the exercise proposed in section 4.4. A fine modeling of the flow must be carried out on a larger area than the *tidal lagoon* itself for quantifying the energy recovered by a marine current turbines farm.

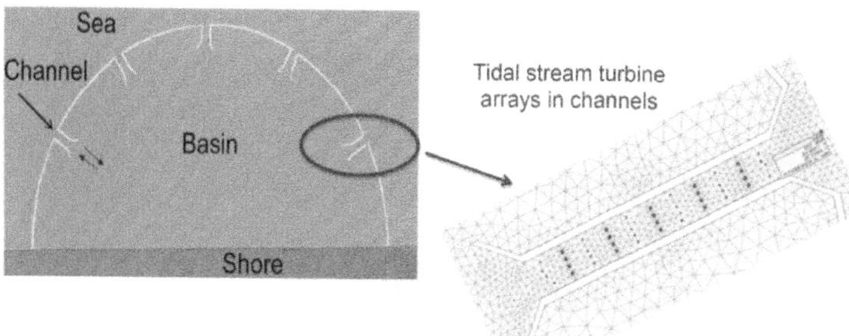

Figure 4.5. *Diagram of a tidal lagoon*[11]

One can wonder whether tidal lagoons and tidal gardens are as different from the classic tidal power plant as claimed. A video[12] promoting a project in Swansea on the River Severn in Wales is worth watching. Admittedly, this project does not block the River Severn with a dam; a basin is isolated on the edge of the river. But it is surprising to see how the turbines system envisaged at Swansea is comparable to that of the Rance power plant. The civil engineering costs are very high since the length (11.5 km) of the dike to be built to create the lagoon is much greater than the length (750 m) of the dam at the Rance. The argument of the project is based on an economic model of shared costs of energy recovery with other activities (sports, education, land use planning). The British government announced in June 2018 its

9 Ibid.

10 Available at: http://www.tidallagoonpower.com/.

11 Source: © EDF/D. Aelbrecht. All rights reserved.

12 Available at: www.tidallagoonpower.com/projects/swansea-bay/3d-model/.

refusal to contribute to the financing of the project, due to its estimated cost in October 2021 at \$2.35 billion.

4.2. Exercise: energy model of the Rance plant

The Rance tidal power plant is based on the concept of closing an estuary with a dam in an area where the amplitude of the tide is high. The dam is equipped with turbines that control the flow of water between the sea and the closed basin of the estuary and recover part of the energy of the flow at rising tide as well as at falling tide. A description of the Rance plant and a visualization of its operation are available on the EDF site[13].

This exercise describes the optimal operation of the plant, which means here that which maximizes the recovered energy. The model is simplified compared to reality: the estuary is schematized as a parallelepiped with ground surface S and the tide is assumed to be sinusoidal with constant amplitude H_{mo}. The water level on the sea side therefore varies according to:

$$H_m(t) = H_0 + \frac{1}{2}H_{mo}\sin(\omega t)$$

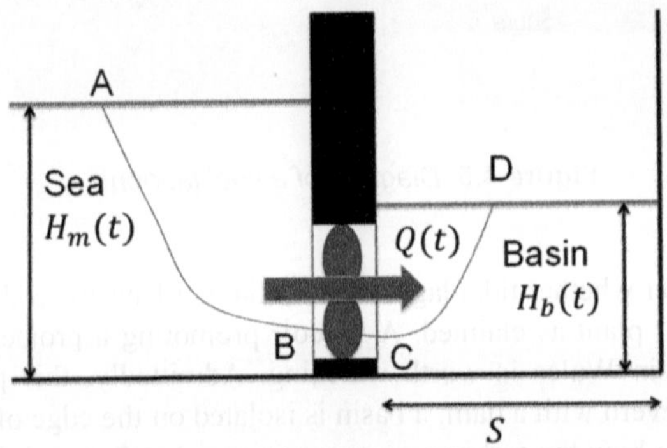

Figure 4.6. *Schematization of the tidal power plant dam equipped with turbines*

H_0 is the average depth of the water layer, H_{mo} is the tidal height and $T = 2\pi/\omega$ is the tidal period.

---

13 Available at: www.edf.fr/groupe-edf/producteur-industriel/energies-renouvelables/energies-marines/energie-des-marees.

We designate by $H_b(t)$ the height of water in the basin, which varies in response to the variation of the sea level and according to what the turbines allow. Figure 4.6 schematizes the device. The ABCD curve is a streamline at the instant represented.

1) Write the mass balance over the volume of the basin to express the flow rate $Q(t)$ passing through the turbines as a function of the water height.

2) For the case of the flow entering the basin, then for that of the outgoing flow, determine the head variation between the inlet and the outlet of the turbine and express the instantaneous power recovered by the turbine. It will by convention be positive when the machine is operating as a turbine and negative when it is in pumping mode.

3) It is assumed that the operation of the turbines controls the flow at any moment of a tidal cycle so that the variation of the water level in the basin is also sinusoidal:

$$H_b(t) = H_0 + \frac{1}{2}H_{b0}\sin(\omega t + \phi)$$

The amplitude H_{b0} in the basin and the phase shift ϕ between the tide at sea and its response in the basin are unknown at this stage. Express the instantaneous power $\wp(t)$ recovered by the turbine for these sinusoidal variations in height at sea and in the basin.

4) Calculate the energy recovered during a tidal period.

5) Determine the conditions that maximize the recovered energy. For these conditions, determine the phase shift ϕ, the amplitude of the water level variations in the basin H_{b0} and the average power recovered.

Plot variations in sea level and water level in the basin over a tidal period, as well as power variations $\wp(t)$.

6) Carry out numerical applications for conditions close to those of the Rance tidal power plant: $S_{turb} = 1{,}200 \text{ m}^2$, basin surface $S = 2{,}000 \text{ ha} = 2 \times 10^7 \text{ m}^2$, tide height $H_{m0} = 10$ m, tidal period $T = 12$ h 25 min. Estimate the amplitude H_{b0}, the average recovered power and the average annual energy recovered.

What environmental impacts do the model results predict?

4.2.1. *Solution*

1) Mass balance on the volume of the basin $SH_b(t)$:

$$\frac{d\big(SH_b(t)\big)}{dt} = Q(t) = S\frac{dH_b(t)}{dt}$$

Positive flow rate is flow entering the basin; it raises the water level.

2) Consider the streamline joining A and D. Neglecting the kinetic energy since the basin and the sea have large volumes, the head at the end points is:

$$H_A = P_{atm} + \rho g H_m(t)$$
$$H_D = P_{atm} + \rho g H_b(t)$$

When the flow enters the basin, the flow is directed from A to D. The head variation between A and D is, applying the generalized Bernoulli equation (section 1.6):

$$H_D - H_A = \rho g H_b(t) - \rho g H_m(t) = -\Delta H_{turb} - \rho\frac{Q^2}{2S_{turb}^2}$$

Since the flow is accelerated between A and B, we assume that there is no head loss between A and B. On the other hand, between C and D, we apply Borda's formula (head loss for a sudden broadening, see section 1.5) to estimate the head loss of the jet exiting the turbines. Assuming that the width of the widened section is large compared to the delivery section S_{turb} at the outlet of the turbine channel, the kinetic energy of the jet is totally lost. Between B and C, the head lost by the fluid is entirely recovered by the turbine (linear head losses are neglected). It follows that:

$$\Delta H_{turb} = -\rho g H_b(t) + \rho g H_m(t) - \rho\frac{Q^2}{2S_{turb}^2}$$

and the instantaneous power recovered by the turbine is:

$$\wp(t) = \Delta H_{turb} Q(t) = \left\{\rho g\big(H_m(t) - H_b(t)\big) - \rho\frac{Q^2}{2S_{turb}^2}\right\} Q(t)$$
$$= \rho g\big(H_m(t) - H_b(t)\big)Q(t) - \rho\frac{Q^3}{2S_{turb}^2}$$

When the flow leaves the basin, the flow is from D to A and the previous equation is modified according to:

$$H_A - H_D = \rho g H_m(t) - \rho g H_b(t) = -\Delta H_{turb} - \rho \frac{Q^2}{2S_{turb}^2}$$

from where:

$$\Delta H_{turb} = -\rho g H_m(t) + \rho g H_b(t) - \rho \frac{Q^2}{2S_{turb}^2}$$

By multiplying by the flow $-Q(t)$ since $Q(t)$ is negative when the flow leaves the basin, the instantaneous power recovered by the turbine is written as:

$$\begin{aligned}\wp(t) = -\Delta H_{turb} Q(t) &= -\left\{\rho g\big(H_b(t) - H_m(t)\big) - \rho \frac{Q^2}{2S_{turb}^2}\right\} Q(t) \\ &= \rho g\big(H_m(t) - H_b(t)\big)Q(t) + \rho \frac{Q^3}{2S_{turb}^2}\end{aligned}$$

An expression of the mechanical power recovered by the turbine, valid for both cases, is:

$$\wp(t) = \rho g\big(H_m(t) - H_b(t)\big)Q(t) - \rho \frac{|Q|^3}{2S_{turb}^2}$$

With the expression of the flow rate obtained in question 1:

$$Q(t) = S\frac{dH_b(t)}{dt}$$

we finally write the instantaneous power recovered by the turbines in the form:

$$\wp(t) = \rho g\big(H_m(t) - H_b(t)\big)S\frac{dH_b(t)}{dt} - \rho \frac{S^3}{2S_{turb}^2}\left|\frac{dH_b(t)}{dt}\right|^3$$

3) By introducing variations in water level in the sea $H_m(t) = H_0 + \frac{1}{2}H_{m0}$ $\sin(\omega t)$ and the response in the basin $H_b(t) = H_0 + \frac{1}{2}H_{b0}sin(\omega t + \phi)$ in the power obtained in the previous question, the mechanical power recovered by the turbines becomes:

$$\wp(t) = \rho g \frac{1}{2}(H_{mo}\sin(\omega t) - H_{bo}\sin(\omega t + \phi))S\frac{1}{2}H_{bo}\omega\cos(\omega t + \phi)$$
$$- \rho \frac{S^3}{2S_{turb}^2}\left(\frac{H_{bo}\omega}{2}\right)^3 |\cos^3(\omega t + \phi)|$$

4) The energy recovered during a tidal period $T = 2\pi/\omega$ is:

$$E = \int_0^T \wp(t)dt$$

$$= \rho g \frac{1}{4} S H_{mo} H_{bo}\omega \int_0^T \sin(\omega t)\cos(\omega t + \phi)dt$$

$$-\rho g \frac{1}{4} S H_{bo}^2 \omega \int_0^T \sin(\omega t + \phi)\cos(\omega t + \phi)dt$$

$$- \rho \frac{S^3}{2S_{turb}^2}\left(\frac{H_{bo}\omega}{2}\right)^3 \int_0^T |cos^3(\omega t + \phi)|dt$$

The mathematical calculations that follow, taking into account what $\omega T = 2\pi$, are, keeping only the non-zero integrals:

$$E = \rho g \frac{1}{4} S H_{mo} H_{bo}\omega \int_0^T \sin(\omega t)\{\cos(\omega t)\cos(\phi) - \sin(\omega t)\sin(\phi)\}dt$$
$$- \rho \frac{S^3}{2S_{turb}^2}\left(\frac{H_{bo}\omega}{2}\right)^3 \frac{1}{\omega}\int_0^{2\pi} |\cos^3 x|dx$$

$$E = -\rho g \frac{1}{4} S H_{mo} H_{bo}\omega\sin(\phi) \int_0^T \sin^2(\omega t)dt$$
$$- \rho \frac{S^3}{2S_{turb}^2}\left(\frac{H_{bo}\omega}{2}\right)^3 \frac{2}{\omega}\int_{-\pi/2}^{\pi/2} \cos^3 x dx$$

$$E = -\rho g \frac{1}{4} S H_{mo} H_{bo}\omega\sin(\phi) \int_0^T \frac{1 - \cos(2\omega t)}{2}dt$$
$$- \rho \frac{S^3}{2S_{turb}^2}\left(\frac{H_{bo}\omega}{2}\right)^3 \frac{2}{\omega}\int_{-\pi/2}^{\pi/2} \cos x(1 - \sin^2 x)dx$$

$$E = -\rho g \frac{1}{4} S H_{m0} H_{b0} \sin(\phi) \frac{\omega T}{2} - \rho \frac{S^3}{2S_{turb}^2} \left(\frac{H_{b0}\omega}{2}\right)^3 \frac{2}{\omega} \left\{ \int_{-\pi/2}^{\pi/2} \cos x dx - \int_{-\pi/2}^{\pi/2} \cos x (\sin^2 x) dx \right\}$$

$$E = -\rho g \frac{1}{4} S H_{m0} H_{b0} \sin(\phi) \pi - \rho \frac{S^3}{2S_{turb}^2} \left(\frac{H_{b0}\omega}{2}\right)^3 \frac{2}{\omega} \left\{ 2 - \int_{-1}^{1} y^2 dy \right\}$$

with $y = \sin x$:

$$E = -\rho g \frac{\pi}{4} S H_{m0} H_{b0} \sin(\phi) - \rho \frac{S^3}{2S_{turb}^2} \left(\frac{H_{b0}\omega}{2}\right)^3 \frac{2}{\omega} \frac{4}{3}$$

Hence, the final relation:

$$E = -\rho g \frac{\pi}{4} S H_{m0} H_{b0} \sin(\phi) - \rho \frac{1}{6} \frac{S^2}{S_{turb}^2} \omega^2 H_{b0}^3 S$$

5) The energy recovered depends on the values of H_{b0} and ϕ that the monitoring of the flow by the turbines imposes. It is maximum when:

$$\frac{dE}{d\phi} = -\rho g \frac{\pi}{4} S H_{m0} H_{b0} \cos(\phi) = 0$$

$$\frac{dE}{dH_{b0}} = -\rho g \frac{\pi}{4} S H_{m0} \sin(\phi) - \rho \frac{1}{2} \frac{S^2}{S_{turb}^2} \omega^2 H_{b0}^2 S = 0$$

The phase shift $\phi = \pi/2$ cannot be retained because H_{b0}^2 must be positive.

On the other hand, for $\phi = -\pi/2$, we deduce:

$$\frac{H_{b0}}{H_{m0}} = \frac{S_{turb}}{S} \frac{1}{\omega} \sqrt{\frac{\pi g}{2H_{m0}}} = \frac{S_{turb}}{S} T \sqrt{\frac{g}{8\pi H_{m0}}}$$

For these optimal operating conditions, the water level variations in the sea and the basin are, respectively:

$$H_m(t) = H_0 + \frac{1}{2} H_{m0} \sin(\omega t)$$

$$H_b(t) = H_0 - \frac{1}{2}H_{b0}\cos(\omega t)$$

They have a time lag of a quarter of period.

The average power recovered during a tide is:

$$\wp_{average} = \frac{E(T)}{T} = \rho g \frac{1}{12} S_{turb} \sqrt{\frac{g}{H_{m0}} \frac{\pi}{2}} H_{m0}^2$$

and the annual energy recovered is in Wh:

$$E_{annual} = \wp_{average} \times 24 \times 365$$

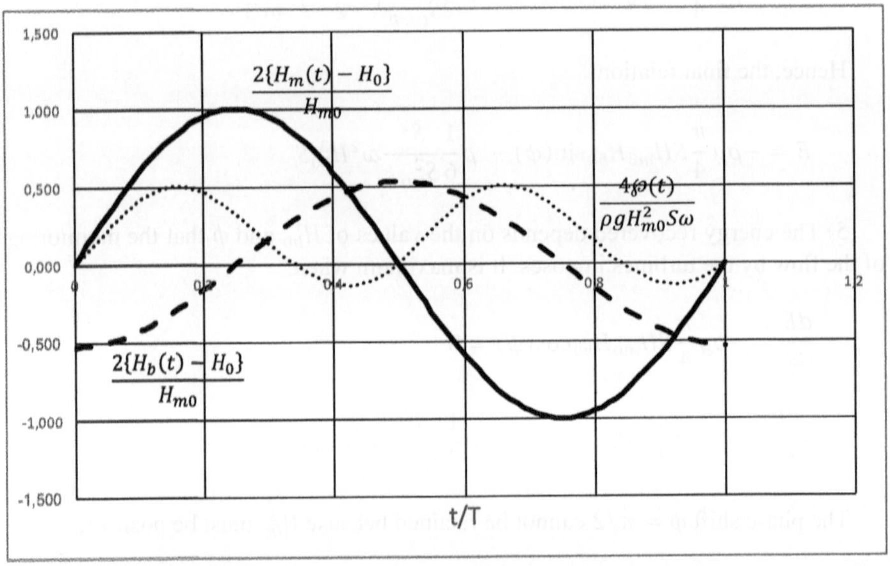

Figure 4.7. *Variations during a tide of the levels at sea and in the basin and of the power recovered by the turbines*

The temporal variations of the water level at sea and in the basin are plotted in Figure 4.7 over the duration of a tide. The temporal variations of the instantaneous power $\wp(t)$ are superimposed on the same graph. The graph highlights the times during a tidal period when the machine is in turbine mode (positive power) and when it is in pump mode (negative power).

6) The numerical applications for the Rance power plant are to be made with the operating conditions:

$S_{turb} = 1{,}200\ \text{m}^2$, basin surface $S = 2{,}000\ \text{ha} = 2 \times 10^7\ \text{m}^2$, tidal height $H_{m0} = 10\ \text{m}$, tidal period $T = 12\ \text{h}\ 25\ \text{min}$.

The calculation carried out by a spreadsheet gives the results presented in Table 4.1.

H_{m0}	T	S_{turb}	S	H_{b0}/H_{m0}	$\wp_{average}$	E_{annual}
M	S	m^2	m^2	–	MW	GWh
10	44,700	1,200	2×10^7	0.530	122	1,067

Table 4.1. *Numerical applications: tidal amplitude in the basin, average power and annual energy recovered by the Rance plant*

The results can be compared with the data provided by EDF[14]:

– the tidal range in the basin is reduced by 40%;

– the installed capacity at the Rance plant is 240 MW;

– the annual energy recovered is 540 GWh.

The model correctly predicts the maximum power (twice the average power) installed on the plant and the reduction in tidal amplitude observed in the basin. On the other hand, the model predicts an annual recovery greater by a factor of 2 than the energy actually recovered. This could be explained by the variations in tidal amplitude that are not considered here and by the fact that the flow control by the turbines could not be as optimal as the model would like. The representations by D. Aelbrecht of the joint evolutions of the water levels at sea and in the basin show a phase shift of about a quarter of a period, in agreement with the model (Figure 4.7). The turbines are indeed used in pumping mode during certain phases of the tide.

A major interest of the model is to explain the reduction in tidal amplitude in the basin. It is caused by the loss of the kinetic energy of the fluid leaving the turbine channel. In this respect, it is important to point out that the surface $S_{turb} = 1{,}200\ \text{m}^2$ retained is greater than the total area of the turbine rotors. The rotor diameter of the Kaplan turbines at the Rance plant is 5.75 m, giving a total rotor area of $540\ \text{m}^2$ for the 24 turbines at the plant. If we had retained this value for S_{turb}, the kinetic energy of the jets and the singular head loss would be much greater. The

14 See Albrecht (2015).

reduction in tidal amplitude in the basin would be greater and the average power recovered would also be lower. This draws attention to the role of the diverging ducts on either side of the rotor up to the ends of the dam, which can be seen on the section of the Rance dam (Figure 4.8). They serve to decrease the kinetic energy. The value $S_{turb} = 1{,}200\ \text{m}^2$ is an estimate of the discharge surface at the entrance and exit of the dam and not at the level of the rotor.

The reduction in tide in the estuary due to the tidal power plant has a significant environmental impact. The estuary is not a parallelepiped as modeled in this exercise. A reduction in the amplitude of the tides means a modification of the intertidal zones (zones alternatively covered at high tide and discovered at low tide) and wetlands, with its consequences for the fauna and flora.

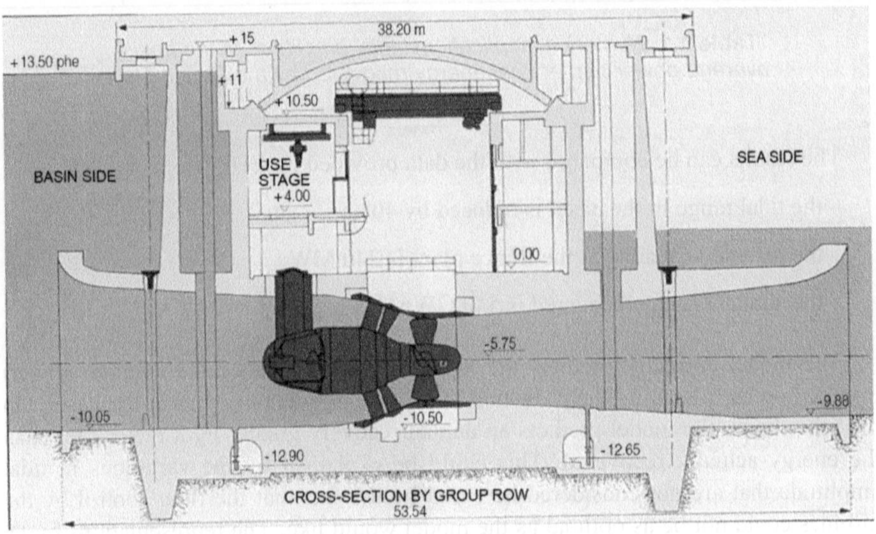

Figure 4.8. *Sectional view of the Rance dam with the installation of a turbine and the divergent channels upstream and downstream*[15]

4.3. Exercise: sizing a Kaplan turbine

A turbine is sized for the conditions required at the Rance power plant: maximum power of 10 MW and flow rate of 275 m^3/s. The radius of the rotor is $R_e = 2.7$ m and the rotor rotates at the speed of 93 rpm. In this exercise, the stator is configured and the rotor blade angles are determined.

15 Source: © EDF. All rights reserved.

At the Rance plant, the angle of the Kaplan turbine blades can be changed over time by rotating the blades, so as to control the flow during a tidal period and to operate at rising and falling tides. The angle of the blades calculated in this exercise is determined for the condition where the flow is maximum.

The schematic of the Kaplan turbine is shown in Figure 4.9. In order for the turbine to operate identically during the ebb and flow, the turbine is equipped with two stators arranged on either side of the rotor. The rotor and stators are placed in a circular cylindrical tube where the flow is cylindrical (hence $u_r = 0$). It is further assumed that the flow is homogeneous in any flow section.

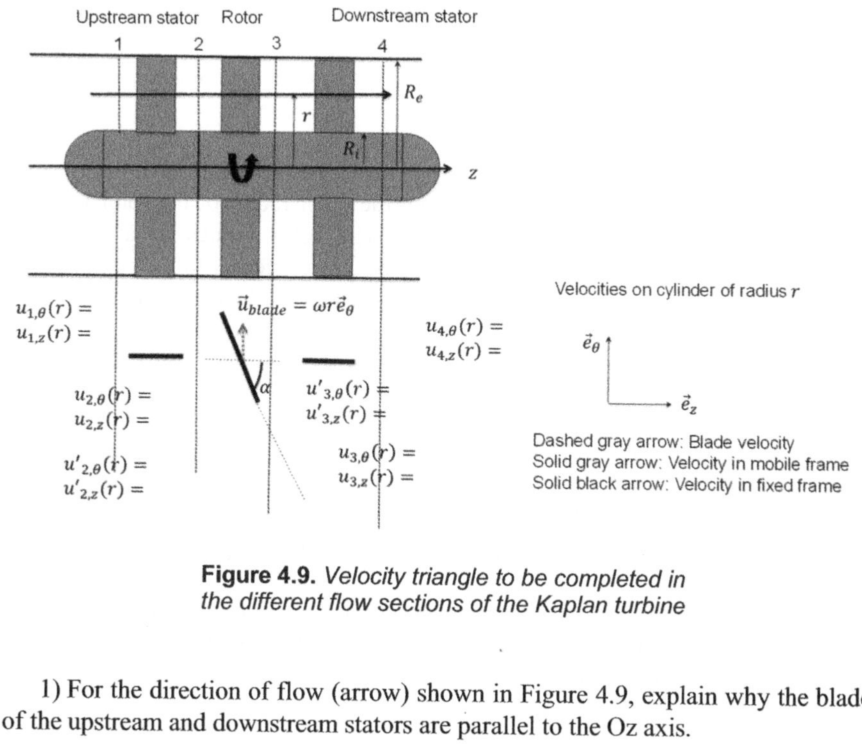

Figure 4.9. *Velocity triangle to be completed in the different flow sections of the Kaplan turbine*

1) For the direction of flow (arrow) shown in Figure 4.9, explain why the blades of the upstream and downstream stators are parallel to the Oz axis.

2) At the radial position r the rotor blades make an angle $\alpha(r)$ with the Oz axis. Turbine sizing is all about finding the right angles to achieve the desired operation. For the direction of the flow indicated in Figure 4.9, express the velocity components (u_θ, u_z) in the fixed reference in the planes 1, 2, 3 and 4. One will also express the velocity components (u'_θ, u'_z) in the reference related to the blades in the planes 2 and 3. Specify what follows from the conservation of flow rate and what follows from the Joukovski condition. Represent the speeds with the code indicated in Figure 4.9.

3) Apply the momentum theorem in the circular annular domain of radius r and thickness dr limited by sections 2 and 3 and deduce the force applied to the rotor element contained in this volume. Explain why velocities in the relative frame are used for this calculation. Deduce the mechanical power recovered by this rotor element.

In this question and up to question 9 (included), the potential energy will be added to the pressure by denoting the piezometric pressure by p.

4) Sizing is done so that all fluid particles give the same head ΔH to the rotor, regardless of the radial position r of the rotor element. Deduce the variations of the blade angle $\alpha(r)$.

5) Check that the velocity field in section 3 is irrotational. Is this an expected result?

6) Each Kaplan turbine can provide a maximum power of 10 MW and a flow rate of 275 m^3/s. As at the Rance power plant, the outer radius of the rotor is $R_e = 2.7$ m and its inner radius is $R_i = 0.675$ m. The rotor rotates at a speed of 93 rpm (source: Wikipedia). Determine the head ΔH transferred to the rotor by the fluid particles and determine the numerical values which fix the blade angles $\alpha(r)$.

7) Calculate the specific speed of the Kaplan turbine rotor and show that it does not depend on its size or on its rotational speed.

8) Determine the angle $\beta_1(r)$ that the flow velocity makes (in the moving frame and the plane 2) with the rotor blade and the angle $\beta_2(r)$ that the flow velocity makes (in the fixed frame and the plane 3) with the blade of the downstream stator. Plot on the same graph the radial variations of angles $\beta_1(r)$, $\beta_2(r)$ and $\alpha(r)$.

By increasing the speed of rotation of the rotor to 200 rpm, plot on the same graph the radial variations of angles $\beta_1(r)$, $\beta_2(r)$ and $\alpha(r)$ for operating the turbine with the same power and the same flow rate. Discuss the effect of changing rotational speed.

9) Deduce from the head ΔH expressed in question 4 the pressure difference $p_3(r) - p_2$ between planes 2 and 3. To do this, the velocities in the fixed frame are used. Calculate the head variation between planes 2 and 3 using the speeds in the moving frame. What does this result mean?

10) Express the absolute pressure at a point in section 2 and at a point in section 3.

It is assumed that, when the tide is operating at maximum power, the axis of the turbine is at a depth $Z_c = -10$ m below sea level. Plot the changes in absolute pressure on a vertical axis passing through the axis of the turbine, in planes 2 and 3, and comment on the layout.

Explain why the risk of cavitation prevents reducing the size of the turbine.

4.3.1. *Solution*

1) The downstream stator removes the rotational component u_θ downstream of the rotor by converting the kinetic energy of this velocity component into pressure. The blade of the downstream stator must therefore be parallel to the axis Oz so as to verify the Joukovski condition at the rear of the stator. Since the direction of the flow is reversible, the upstream and downstream stators are identical. The upstream stator plays no role because the flow is parallel to the upstream blade. But when the direction of the flow is reversed, the upstream stator becomes the downstream stator and vice versa.

2) The flow section is the same in all planes. As the flow rate is assumed to be homogeneous, the component $u_z(r)$ has the same value in all planes:

$$U = \frac{Q}{\pi(R_e^2 - R_i^2)}$$

In each plane, the speed in the fixed frame and the speed in the moving frame are linked by the composition of the velocities:

$$\vec{u}(r) = \overrightarrow{u'}(r) + \vec{u}_{blade} = \overrightarrow{u'}(r) + \omega r \vec{e}_\theta$$

The Joukovski condition is used to express the velocity in the relative frame in section 3, knowing its axial component.

The velocity triangle is shown in Figure 4.10. In planes 1, 2 and 4, the velocities in the fixed frame are equal to the axial velocity component.

The conversion processes at the tidal power plant are also summarized in the upper part of Figure 4.10. In this regard, the dam is associated with the Kaplan turbine by converting the potential energy upstream into pressure energy at the turbine inlet, as a penstock supplying a Pelton turbine or a Francis turbine in a hydroelectric installation does. For the direction of the flow oriented by the arrow passing through the turbine (Figure 4.9), the upstream stator plays no role in terms of energy conversion. The second conversion step is carried out by the rotor which, coupled to the alternator, converts the pressure energy into electrical energy. The rotor also generates rotational flow in the fluid, incidentally converting pressure energy into kinetic energy in the fluid. The third conversion step is ensured by the downstream stator, which converts the rotational kinetic energy produced by the rotor into pressure.

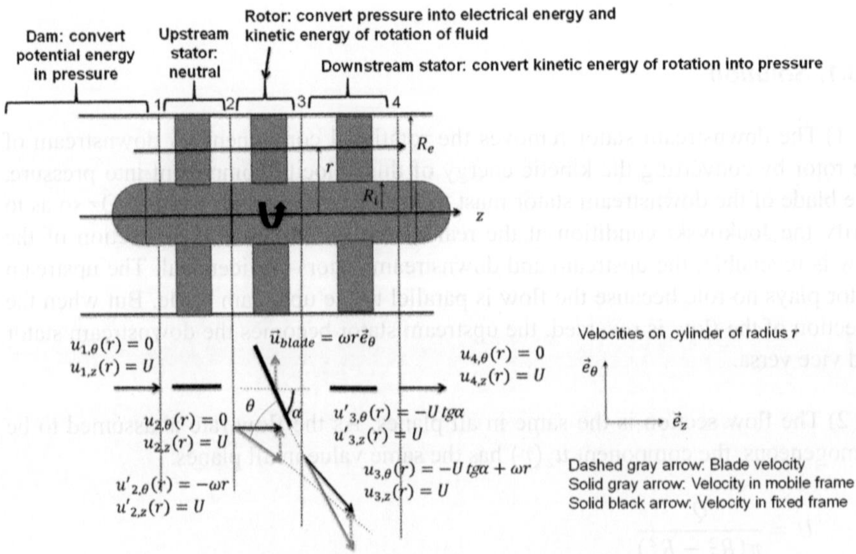

Figure 4.10. *Velocity triangle completed in the different flow sections of the Kaplan turbine*

3) The momentum theorem (section 1.8) requires the flow to be permanent. It is therefore necessary to place yourself in the reference mark linked to the blades to apply it. The force applied to the rotor element contained in the annular cylinder of radius r and thickness dr between the delivery sections 2 and 3 is calculated:

$$\overrightarrow{dF} = -\left\{\rho\left(\overrightarrow{u'}_2.(-\vec{e}_z)\right)\overrightarrow{u'}_2 + p_2(-\vec{e}_z)\right\}(2\pi r dr)$$
$$-\left\{\rho\left(\overrightarrow{u'}_3.(\vec{e}_z)\right)\overrightarrow{u'}_3 + p_3(\vec{e}_z)\right\}(2\pi r dr)$$

Since the potential energy is added to the pressure, p_2 and p_3 are the piezometric pressures, which depend only on r.

By introducing the speeds indicated in Figure 4.10, we deduce:

$$\overrightarrow{dF} = \rho U(Utg\alpha - \omega r)\vec{e}_\theta(2\pi r dr) + (p_2 - p_3)\vec{e}_z(2\pi r dr)$$

The power recovered by this rotor element is therefore:

$$d\wp = \overrightarrow{dF}.(\omega r\vec{e}_\theta) = \rho U(Utg\alpha - \omega r)\omega r(2\pi r dr) = \rho(Utg\alpha - \omega r)\omega r dQ$$

where $dQ = U(2\pi r dr)$ is the flow rate passing through the annular cylinder of radius r and thickness dr.

4) Since $d\wp = \rho(Utg\alpha - \omega r)\omega r dQ = \Delta H dQ$, it follows that:

$$tg\{\alpha(r)\} = \frac{\Delta H}{\rho\omega U R_e}\frac{R_e}{r} + \frac{\omega R_e}{U}\frac{r}{R_e} = a\frac{R_e}{r} + b\frac{r}{R_e}$$

If ΔH is a constant (does not depend on r), then the dimensionless constants:

$$a = \frac{\Delta H}{\rho\omega U R_e} \quad \text{and} \quad b = \frac{\omega R_e}{U}$$

fix the variation of the blade angle with r.

5) In plane 3, the speed in the fixed frame is written, using the results of question 4:

$$u_{3,r} = 0,\ u_{3,\theta} = -Utg\alpha + \omega r = -\frac{\Delta H}{\rho\omega}\frac{1}{r},\ u_{3,z} = U$$

It is verified that the rotational of the velocity is zero. It is written in the cylindrical coordinate system:

$$\overrightarrow{Rot\vec{u}} = \begin{pmatrix} \frac{1}{r}\frac{\partial u_z}{\partial\theta} - \frac{\partial u_\theta}{\partial z} \\ \frac{\partial u_r}{\partial z} - \frac{\partial u_z}{\partial r} \\ \frac{1}{r}\frac{\partial(r u_\theta)}{\partial r} - \frac{1}{r}\frac{\partial u_r}{\partial\theta} \end{pmatrix} = \vec{0}$$

This property indicates the proper functioning of the rotor. It avoids fluid head loss other than the head recovered by the rotor. The flow is irrotational upstream of the rotor and remains so downstream.

6) Power and head variation are related by:

$$\wp = \Delta H.Q$$

Numerical values are shown in Table 4.2.

$\wp$	Q	ΔH	R_e	R_i	n	ω	U	a	b
MW	m³/s	Pa	M	m	Rpm	rad/s	m/s	–	–
10	275	36,364	2.7	0.675	93	9.74	12.81	0.108	2,053

Table 4.2. *Numerical applications: flow rate velocity U in the turbine and parameters a and b determining the angle of the blades with the axis Oz*

7) Calculation of the specific speed using flow rate and head variation:

$$n_s = n\frac{Q^{1/2}}{\left(\frac{\Delta H}{\rho g}\right)^{3/4}} = 577$$

Using the expressions for the coefficients a and b obtained in question 4:

$$n_s = n\frac{Q^{1/2}}{\left(\frac{\Delta H}{\rho g}\right)^{3/4}} = \frac{60\omega}{2\pi}\frac{\left(\pi U(R_e^2 - R_i^2)\right)^{1/2}}{\left(\frac{a\omega R_e U}{g}\right)^{3/4}} = \frac{60\omega}{2\pi}\frac{\left(\pi\frac{\omega R_e}{b}(R_e^2 - R_i^2)\right)^{1/2}}{\left(\frac{a\omega R_e \omega R_e}{gb}\right)^{3/4}}$$

Hence:

$$n_s = \frac{30g^{3/4}}{\sqrt{\pi}}\frac{b^{1/4}}{a^{3/4}}\left(1 - \frac{R_i^2}{R_e^2}\right)^{1/2}$$

The specific speed depends only on the dimensionless values a and b, which determine the variations of the blade angle between the hub and the blade tip, and on the ratio R_i/R_e. On the other hand, it does not depend on the angular speed of rotation ω of the rotor, nor on the size of the rotor. If the external radius of the rotor R_e is increased, the radius R_i of the hub is increased in the same proportions to maintain the geometry of the rotor.

8) The tangent unit vector to the rotor blade is $\vec{t}_{blade,rot} = -\sin\alpha\vec{e}_\theta + \cos\alpha\vec{e}_z$.

The unit vector parallel to the relative velocity of the flow with respect to the blade (velocity in the rotating frame), in plane 2, is:

$$\vec{t'}_{vel,2} = \frac{-\omega r}{\sqrt{U^2 + \omega^2 r^2}}\vec{e}_\theta + \frac{U}{\sqrt{U^2 + \omega^2 r^2}}\vec{e}_z$$

The angle of incidence of the flow with respect to the blade is therefore:

$$\cos\beta_1 = \vec{t}_{blade,rot}.\vec{t'}_{vel,2} = \cos\alpha\frac{\omega r tg\alpha + U}{\sqrt{U^2+\omega^2 r^2}} = \frac{1}{\sqrt{1+tg^2\alpha}}\frac{\omega r tg\alpha + U}{\sqrt{U^2+\omega^2 r^2}}$$

The unit vector tangent to the stator blade is $\vec{t}_{blade,stat} = \vec{e}_z$.

Let us recall the velocity in the fixed frame in plane 3 (Figure 4.10):

$$u_{3,r} = 0,\ u_{3,\theta} = -Utg\alpha + \omega r = -\frac{\Delta H}{\rho\omega}\frac{1}{r},\ u_{3,z} = U$$

The unit vector parallel to the relative velocity of the flow with respect to the stator blade (velocity in the fixed frame), in plane 3, is then:

$$\vec{t}_{vel,3} = -\frac{\frac{\Delta H}{\rho\omega}\frac{1}{r}}{\sqrt{U^2+\left(\frac{\Delta H}{\rho\omega}\frac{1}{r}\right)^2}}\vec{e}_\theta + \frac{U}{\sqrt{U^2+\left(\frac{\Delta H}{\rho\omega}\frac{1}{r}\right)^2}}\vec{e}_z$$

The angle of incidence of the flow with respect to the blade of the downstream stator is therefore:

$$\cos\beta_2 = \vec{t}_{blade,stat}.\vec{t}_{vel,3} = \frac{U}{\sqrt{U^2+\left(\frac{\Delta H}{\rho\omega}\frac{1}{r}\right)^2}}$$

Figure 4.11 shows the variation of the angle $\alpha(r)$ of the rotor blades with the Oz axis as well as the angles of incidence of the flow with respect to the rotor blades and the downstream stator blades. The angle $\alpha(r)$ is close to 45° and increases with r, which is in agreement with the images of the rotor of the Kaplan turbine at the Rance plant.

The angles of incidence are small enough for it to be reasonable to think that flow separation is avoided. Difficulties in this respect appear when approaching the hub. This perhaps explains why the radius R_i of the hub is quite large, because the operation of the turbine is less good on this side, the risk of flow separation becoming significant.

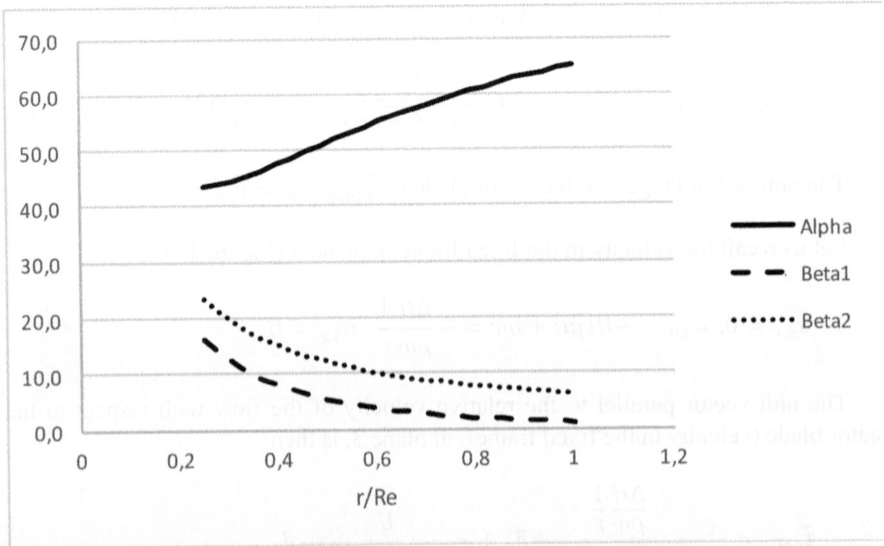

Figure 4.11. *Variations of rotor blade angle $\alpha(r)$, of flow angle $\beta_1(r)$ with rotor blades (velocities in moving frame), and flow angle $\beta_2(r)$ with stator blades (velocities in fixed frame), for the angular speed of rotation $n = 93\ rpm$*

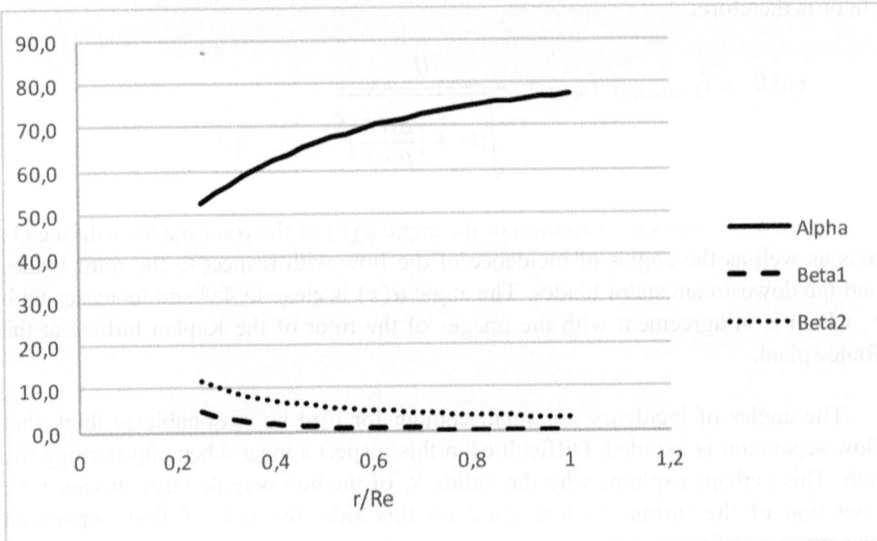

Figure 4.12. *Variations of rotor blade angle $\alpha(r)$, of flow angle $\beta_1(r)$ with rotor blades (velocities in moving frame), and flow angle $\beta_2(r)$ with stator blades (velocities in fixed frame), for the angular speed of rotation $n = 200\ rpm$*

If the speed of rotation is increased to 200 rpm (Figure 4.12), the angle $\alpha(r)$ becomes greater. The rotor blades are more perpendicular to the direction of the delivering flow. The angles of incidence relative to the rotor and stator blades are reduced.

9) The piezometric pressure p_2 is uniform in plane 2 because the head and velocity components are both uniform. In plane 3, as $u_{3,\theta}$ varies with r, the piezometric pressure p_3 also varies with r:

$$p_3(r) - p_2 = -\Delta H + \rho\frac{u_{2,\theta}^2}{2} + \rho\frac{u_{2,z}^2}{2} - \rho\frac{u_{3,\theta}^2}{2} - \rho\frac{u_{3,z}^2}{2}$$
$$= -\Delta H - \rho\frac{1}{2}\left(\frac{\Delta H}{\rho\omega}\frac{1}{r}\right)^2$$

The head variation between sections 2 and 3 when the speeds taken into account are those in the relative reference is written as:

$$\Delta H' = p_3(r) + \rho\frac{u'^2_{3,\theta}}{2} + \rho\frac{u'^2_{3,z}}{2} - p_2 - \rho\frac{u'^2_{2,\theta}}{2} - \rho\frac{u'^2_{2,z}}{2}$$
$$= -\Delta H - \rho\frac{1}{2}\left(\frac{\Delta H}{\rho\omega}\frac{1}{r}\right)^2 + \rho\frac{U^2 tg^2\alpha}{2} - \rho\frac{\omega^2 r^2}{2}$$

By expressing the variations of $tg\{\alpha(r)\}$ using the results of question 4:

$$\Delta H' = -\Delta H - \rho\frac{1}{2}\left(\frac{\Delta H}{\rho\omega}\frac{1}{r}\right)^2 + \rho\frac{U^2}{2}\left(\frac{\Delta H}{\rho\omega r U} + \frac{\omega r}{U}\right)^2 - \rho\frac{\omega^2 r^2}{2} = 0$$

In the frame linked to the blades, the flow is stationary. The result $\Delta H' = 0$ means that the flow is ideal. It does not undergo any head loss when the speeds taken into account are those in the moving frame.

When the head change is calculated using the velocities in the fixed frame, the flow yields the head ΔH recovered entirely by the rotor.

10) Since there is no head variation between the free surface in the sea and plane 2, the piezometric pressure p_2 in plane 2 is:

$$p_2 = P_{atm} - \rho\frac{u_{2,\theta}^2}{2} - \rho\frac{u_{2,z}^2}{2} = P_{atm} - \rho\frac{U^2}{2}$$

The absolute pressure $P_2 = p_2 - \rho g Z$ in plane 2 at altitude Z is then:

$$P_2(r,Z) = P_{atm} - \rho\frac{U^2}{2} - \rho g Z$$

Taking into account the head ΔH transferred to the turbine, we also have, for the piezometric pressure in plane 3:

$$p_3(r) = P_{atm} - \rho\frac{u_{3,\theta}^2}{2} - \rho\frac{u_{3,z}^2}{2} - \Delta H$$

The absolute pressure in plane 3 at altitude Z is then:

$$P_3(r,Z) = P_{atm} - \Delta H - \rho\frac{U^2}{2} - \rho\frac{1}{2}\left(\frac{\Delta H}{\rho\omega}\frac{1}{r}\right)^2 - \rho g Z$$

Z is the altitude of the point along an axis oriented in the opposite direction to gravity (not to be confused with the z direction of the turbine axis) and $Z = 0$ is taken at the sea level.

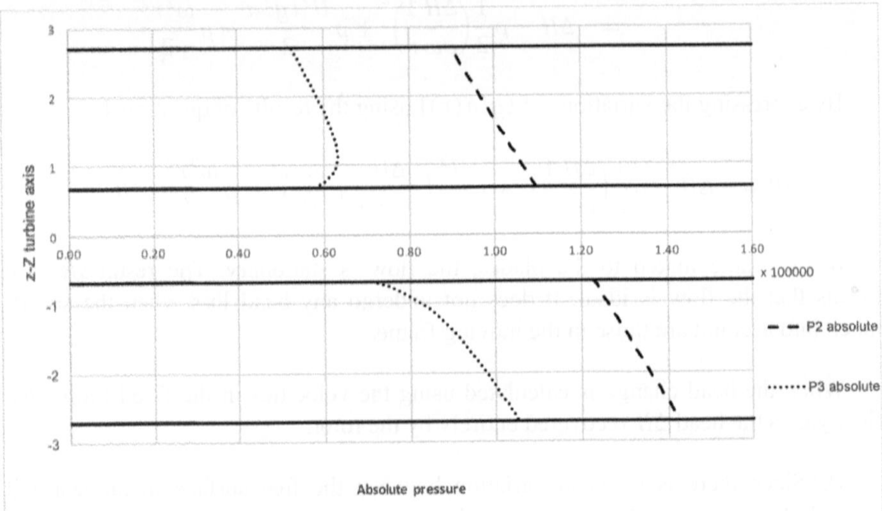

Figure 4.13. *Variation of absolute pressure on a vertical axis passing through the axis of rotation of the Kaplan turbine for the angular velocity of rotation* $n = 93\ rpm$

The variations of the absolute pressure on a vertical axis passing through the axis of the turbine, in planes 2 and 3, are plotted in Figure 4.13 for positions in the intervals $Z_c - R_e < Z < Z_c - R_i$ and $Z_c + R_i < Z < Z_c + R_e$ when the axis of the turbine is at depth $Z_c = -10$ m below sea level.

The absolute pressure profile in plane 2 shows the decrease in pressure with height due to gravity. Throughput speed $U = 12.81\ \mathrm{m/s}$ is important. The pressure drop that results from the acceleration of the fluid in the pipe is compensated by the depth. A decrease in the size of the turbine would increase the speed U and decrease the pressure in this section. The risk of cavitation prevents reducing the radius R_e of the turbine.

The absolute pressure in plane 3 is reduced compared to plane 2 due to the head recovered by the turbine. The decrease near the axis due to the kinetic energy of the component $u_{3,\theta}$ remains limited.

4.4. Exercise: marine current turbines farm and modification of currents

A marine current turbine recovers part of the kinetic energy of a fluid passing through it. In this exercise, we consider a turbine with a diameter of $D = 15$ m. The surface swept by the blades is $S = \pi D^2/4$. When this turbine is placed in a uniform flow with speed U perpendicular to the plane of the rotor (space is of infinite dimensions), the power recovered by the turbine is:

$$\wp_{turbine} = c_p \rho \frac{U^3 S}{2}$$

The power coefficient, measured experimentally, is $c_p = 0.4$. This value is lower than the maximum value $16/27$ predicted by Betz's formula (see exercise 3.2).

It is planned to install a marine current turbines farm at sea, shown schematically in Figure 4.14. It is equipped with 120 turbines arranged in six columns of $N = 20$ machines. The figure only shows part of the farm, with two columns of five turbines. In the absence of marine current turbines, the flow speed $U_{sea} = 2\ \mathrm{m/s}$ is uniform according to the direction Ox throughout the domain. The width of the right-of-way of each column of marine current turbines is $l = 10D = 150$ m and the spacing according to the direction of the flow between two successive turbines is $L = 40D = 600$ m. The latter is assumed to be sufficient to allow for the flow to have become homogeneous again in a plane (x,y) upstream of the next row of turbines. The sections where the flow is homogeneous are shown in dotted lines in Figure 4.14. The lateral spacing in the direction Oy between the axes of rotation of two neighboring turbines being $l = 10D$, it is assumed that a lateral confinement effect can also be ignored. This amounts to assuming that the marine current turbines placed in the farm operate with the value of the power coefficient $c_p = 0.4$ measured for a single marine current turbine in an infinite medium.

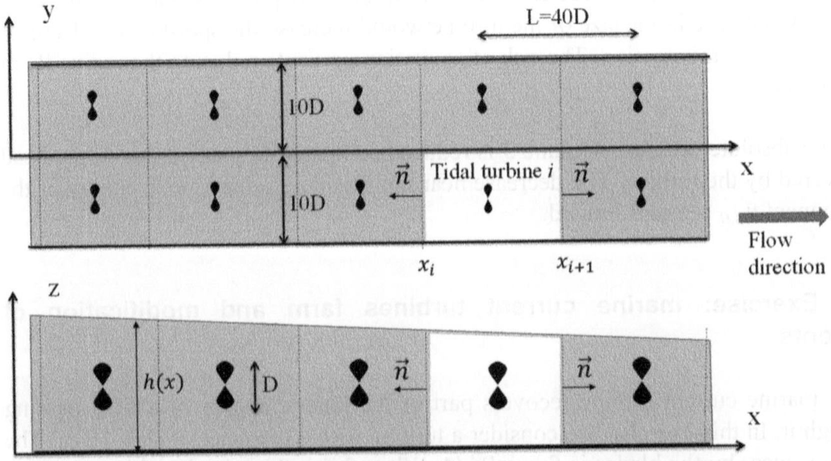

Figure 4.14. *Plan view of a marine current turbines farm confined in a canal (the dimensions of the turbines are not to scale)*

1) The assumptions on the spacing of the turbines suggest that each turbine in the farm collects energy from the current like an isolated turbine in a uniform flow of speed $U_{sea} = 2$ m/s. For these current conditions, determine the power recovered by a marine current turbine as well as the power recovered by the farm.

2) The purpose of this exercise is to study how a marine current turbine farm modifies the currents and thus reduces the power recovered by the turbines. We therefore aim at determining how the speed $U(x_i)$ varies from one row to the next. The coordinate x_i designates the positions in Figure 4.14 of the dotted planes in which the flow is homogeneous. The six columns of the marine current turbines farm are enclosed laterally between two dykes. In reality, we do not plan to build such dykes at sea. In this exercise, they represent a natural channel limited on one side by the coast and on the other side by reefs, which force the flow along the Ox direction. It is an idealized representation of the Alderney Race (Figure 4.2), at the western tip of the Cotentin, where the tidal currents are intense.

The kinetic energy balance on a volume V_i enclosing a marine current turbine is written as:

$$\frac{\partial}{\partial t}\iiint_{V_i}\left\{\frac{1}{2}\rho\left(\vec{u}.\vec{u}\right)+\frac{1}{2}\rho u_{rms}^2\right\}dv$$
$$=-\iint_{S_i+S_{i+1}}\left\{\overline{H}+\frac{1}{2}\rho u_{rms}^2\right\}\left(\vec{u}.\vec{n}\right)ds-c_p\frac{1}{2}\rho U_i^3 S$$
$$-\iiint_{V_i}\rho\varepsilon dv$$

Compare this energy balance with that presented in section 1.7 and explain the meaning of each term. The domain V_i is the domain in white in Figure 4.14, containing the marine current turbine i, bounded by the incoming flow section S_i at $x = x_i$ and the outgoing flow section S_{i+1} at $x = x_{i+1}$. The unit vector $\vec{n}$ on the surfaces S_i and S_{i+1} is by convention oriented outwards from the domain V_i . The speed in the section S_i upstream of the marine current turbine is noted U_i. The flow is stationary. Its parameters (average velocity, head and turbulent velocity) are homogeneous in the planes S_i and S_{i+1}. Let us write the simplifications in the kinetic energy balance.

3) The flow in the sea is a free surface flow. The pressure at the sea surface is atmospheric pressure P_{atm}. The bottom of the sea is flat and the thickness of the water layer is $h_0 = 40$ m when the ocean is at rest.

For free surface flow, it is customary to describe the flow by considering the streamline on the sea surface, which is called the water line. What is the point? Why does water depth h_i vary when velocity also varies U_i in space?

Write the kinetic energy balance by expressing the head as a function of water depth and velocity.

4) In this question and the next, turbulence is neglected. The rate of dissipation of kinetic energy ε and the turbulent velocity u_{rms} are taken equal to 0. Write the simplified kinetic energy balance.

Furthermore, we now consider the channel of the marine current turbines farm in an infinite space (Figure 4.15). In the absence of turbines in the channel, the flow is uniform throughout the space. As before, its speed along the direction Ox is $U_{sea} = 2$ m/s and the depth is h_0 in the entire fluid domain. On the other hand, when marine current turbines are installed, the flow in the channel is reduced. The speeds U_i in the various sections S_i are unknowns of the problem which will have to be determined.

The flow is fluvial when $U < \sqrt{gh}$. Conversely, it is said to be torrential when $U > \sqrt{gh}$. In the present case, the flow is fluvial because the Froude number value:

$$Fr = \frac{U_{sea}^2}{gh_0} = 10^{-2}$$

is very small ($U_{sea} = 2$ m/s and $h_0 = 40$ m).

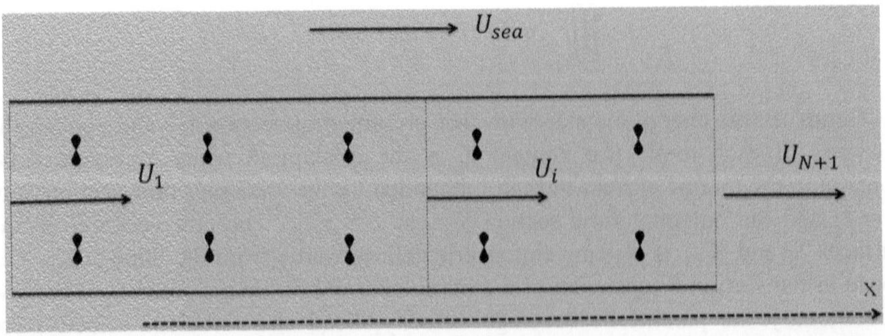

Figure 4.15. *Channel of a marine current turbines farm in an infinite marine space*

In fluvial flow conditions, show that the variations in water depth in the farm are very small compared to the value of $h_0 = 40$ m, that is:

$$\frac{(h_i - h_{i+1})}{h_0} \ll 1$$

Show that the speed variations in the farm are also very small.

5) To determine the variations in water height and speed in the marine current turbines farm, we introduce the expressions:

$$h_i = h_0 + h'_i \text{ and } U_i = U_{N+1} + U'_i$$

Introduce these expressions in the flow rate conservation equations over the sections S_i and in the kinetic energy balances on the volumes V_i.

We pose as an additional condition that the thickness of the water layer at the outlet of the channel is $h_{N+1} = h_0$. What is the physical interpretation of this condition?

Linearize then the different expressions to keep only the first-order terms of the variables h'_i and U'_i.

6) To solve the problem, we must determine the $(N+1)$ unknowns h'_i and the $(N+1)$ unknowns U'_i. Summarize the equations that allow the resolution using the conservation of flow rate and the balances of kinetic energy.

Let us express a final condition at the entrance to the turbines farm channel.

From this set of equations, determine successively U_{N+1}, h'_i and U'_i for i ranging from 1 to N. Finally, analytical expressions will be established and numerical applications will be made.

Estimate the power recovered by the marine current turbines farm and compare with the value estimated in question 1 and comment on the results.

7) Returning to the kinetic energy balance introduced in question 2, in which the terms relating to turbulence will be kept, discuss how turbulence modifies the results obtained in question 6.

4.4.1. *Solution*

1) The power recovered by a marine current turbine with a diameter $D = 15$ m in the current $U_{sea} = 2\ \mathrm{m/s}$ is calculated with the relationship:

$$\wp_{turbine} = c_p \rho \frac{U_{sea}^3 S}{2}$$

The power recovered by the farm is determined by multiplying this result by the number of marine current turbines.

D	S	c_p	U_{sea}	$\wp_{turbine}$	Number of marine current turbines	$\wp_{farm}$
m	m^2	–	m/s	MW	–	MW
15	176.7	0.4	2	0.283	120	33.9

Table 4.3. *Powers recovered by a marine current turbine and by the farm estimated when neglecting a modification of current by the farm*

2) The kinetic energy balance on a volume V_i containing a marine current turbine is written as:

$$\frac{\partial}{\partial t}\iiint_{V_i}\left\{\frac{1}{2}\rho(\vec{u}.\vec{u})+\frac{1}{2}\rho u_{rms}^2\right\}dv$$
$$=-\iint_{S_i+S_{i+1}}\left\{\overline{H}+\frac{1}{2}\rho u_{rms}^2\right\}(\vec{u}.\vec{n})ds - c_p\frac{1}{2}\rho U_i^3 S$$
$$-\iiint_{V_i}\rho\varepsilon dv$$

This equation is similar to the balance written in section 1.7, except for the power recovered by the marine current turbine (second term to the right of the equation), which is new.

The term on the left of the equality is the temporal variation of the kinetic energy (mean flow and turbulence) contained in the domain. In steady state, this term is zero.

The first term to the right of the equality is the flow rate of head and turbulent kinetic energy on the incoming section S_i minus the flow rate of the same quantities on the outgoing section S_{i+1}.

The third term to the right of the equality is the viscous dissipation of kinetic energy in the volume. This term is negative.

When the head and the turbulent kinetic energy are homogeneous on sections S_i and S_{i+1}, these two quantities can be taken out of the integral. When in addition the flow is permanent, the balance is simplified according to:

$$Q\left\{\overline{H}_i+\frac{1}{2}\rho u_{rms,i}^2\right\}-Q\left\{\overline{H}_{i+1}+\frac{1}{2}\rho u_{rms,i+1}^2\right\}=c_p\frac{1}{2}\rho U_i^3 S+\iiint_{V_i}\rho\varepsilon dv$$

The term on the left of the equality is positive, as are each of the two terms on the right of the equality. On the left is the available energy flow, which will be distributed between the power recovered by the marine current turbine contained in the volume and the dissipation of kinetic energy in the volume.

Q denotes the flow rate passing through the sections S_i and S_{i+1}. The index is omitted because the flow rate is conserved when the flow is permanent. It therefore has the same value for all sections S_i.

3) The free surface is a streamline. At the free surface, the pressure is constant and equal to atmospheric pressure. The head is therefore written as:

$$H = P_{atm} + \rho g h + \rho \frac{U^2}{2}$$

It is convenient to use the water line as the stream line of the flow. Its position is determined by the height above the horizontal background. By considering the water line, it is seen that pressure plays no role in the dynamics of the flow. The variation of the head is governed solely by variations in speed U and depth h.

For stationary flow, the flow rate must be maintained between all sections S_i:

$$Q = 10DU_i h_i$$

Therefore, h_i must decrease if U_i increases and vice versa.

By replacing the head by its expression, the kinetic energy balance becomes:

$$Q\left\{\rho g(h_i - h_{i+1}) + \frac{1}{2}\rho(U_i^2 - U_{i+1}^2) + \frac{1}{2}\rho\left(u_{rms,i}^2 - u_{rms,i+1}^2\right)\right\}$$
$$= c_p \frac{1}{2}\rho U_i^3 S + \iiint_{V_i} \rho \varepsilon dv$$

A digression concludes this question. The similarity of the energy conversion processes by wind turbine and by a marine current turbine was mentioned in Chapter 3. For a wind turbine, three successive conversion processes have been schematized in Figure 3.6: a conversion of kinetic energy from the wind into pressure energy upstream of the wind turbine, a conversion of pressure energy into electrical energy as the fluid particles pass through the rotor, and then again a conversion of kinetic energy into pressure energy downstream of the wind turbine.

It is said earlier in this question that pressure plays no role in free surface flow. Therefore, it is difficult to understand how the conversion mechanisms could be identical for a wind turbine and for a marine current turbine at sea. In the present question, the role of the vertical displacement of the free surface is highlighted. A rise in the free surface induces an increase in pressure in the water column by hydrostatic effect. It must therefore be understood that an overpressure in the water column is produced upstream of the marine current turbine by the elevation of the free surface, while downstream there is a lowering of the water level with respect to the medium level. The similarity of energy conversion processes is therefore very real.

4) Neglecting the turbulence, the kinetic energy balance is simplified:

$$Q\left\{g(h_i - h_{i+1}) + \frac{1}{2}(U_i^2 - U_{i+1}^2)\right\} = c_p\frac{1}{2}U_i^3 S$$

With the expression of the flow rate $Q = 10DU_i h_i$, it becomes:

$$2\frac{(h_i - h_{i+1})}{h_0} + \frac{(U_i^2 - U_{i+1}^2)}{U_{sea}^2}\frac{U_{sea}^2}{gh_0} = c_p\frac{S}{h_0 10D}\frac{U_i^2}{U_{sea}^2}\frac{U_{sea}^2}{gh_0}\frac{h_0}{h_i}$$

Suppose, as we will verify later, that $U_i < U_{sea}$. As $Fr = U_{sea}^2/gh_0 = 10^{-2}$, the flow is a fluvial flow. We deduce from the equation above that:

$$\frac{(h_i - h_{i+1})}{h_0} \ll 1$$

The depth of water in the channel therefore remains very close to the depth h_0.

As finally the flow rate $Q = 10DU_i h_i$ is preserved between all sections S_i inside the farm, we deduce that the values of U_i vary as little in the marine current turbine farm.

5) We introduce:

$$h_i = h_0 + h'_i$$

$$U_i = U_{N+1} + U'_i$$

The speed U_{N+1} which is the speed at the exit of the channel is chosen as reference, implying that $U'_{N+1} = 0$.

Likewise, $h'_{N+1} = 0$ since $h_{N+1} = h_0$. The water level is the same on either side of the wall at the outlet of the turbine channel. There is therefore no pressure gradient in the horizontal plane (x, y) at the rear of the farm. This condition means that the flow is a parallel flow at the outlet of the marine current turbines farm (see section 1.3).

The conservation of flow rate between all sections S_i is written as:

$$Q = 10DU_i h_i = 10D(U_{N+1} + U'_i)(h_0 + h'_i) = 10DU_{N+1}h_0$$

Hence the relation, to the first order:

$$U_{N+1}h'_i + U'_i h_0 = 0$$

which connects the height and speed variations in the channel.

The kinetic energy balance on the volume V_i also simplifies to the first order of the variables h'_i and U'_i:

$$g(h'_i - h'_{i+1}) + U_{N+1}(U'_i - U'_{i+1}) = \frac{c_p S}{20D}\frac{(U_{N+1} + U'_i)^2}{h_0 + h'_i}$$
$$\cong \frac{c_p S}{20D}\frac{U_{N+1}^2}{h_0}\left\{1 + 2\frac{U'_i}{U_{N+1}} - \frac{h'_i}{h_0}\right\}$$

With the relation between h'_i and U'_i drawn from the discharge, the N kinetic energy balance equations are transformed into relations involving solely the heights h'_i:

$$\left\{1 - \frac{U_{N+1}^2}{gh_0}\right\}\frac{(h'_i - h'_{i+1})}{h_0} = \frac{c_p S}{20D}\frac{U_{N+1}^2}{gh_0}\left\{1 - 3\frac{h'_i}{h_0}\right\} \quad [4.1]$$

6) The N flow rate conservation equations are written for i ranging from 1 to N:

$$U_{N+1}h'_i + U'_i h_0 = 0 \quad [4.2]$$

We therefore have $2N$ equations considering flow rate conservation ([4.1]) and energy balance ([4.2]) to determine $N + 1$ heights h'_i and $N + 1$ velocities U'_i for i ranging from 1 to $N + 1$.

In the output section S_{N+1}, we additionally know that $h_{N+1} = h_0$, which implies that $h'_{N+1} = 0$. There remains an unknown, the velocity U_{N+1}, for which an additional equation must be found. This relation must establish the upstream link with the incident flow conditions. It is assumed here that the head is conserved between infinity upstream and section S_1. The assumption of using Bernoulli's theorem here is justified by considering that the changes in current are very gradual and therefore do not cause any head loss. We therefore write:

$$\rho g h_0 + \rho\frac{U_{sea}^2}{2} = \rho g(h_0 + h'_1) + \rho\frac{(U_{N+1} + U'_1)^2}{2}$$
$$\cong \rho g(h_0 + h'_1) + \rho\frac{U_{N+1}^2}{2} + \rho U_{N+1}U'_1$$
$$= \rho g(h_0 + h'_1) + \rho\frac{U_{N+1}^2}{2} - \rho U_{N+1}^2\frac{h'_1}{h_0}$$

Or to first order:

$$\frac{U_{sea}^2}{2} = \frac{U_{N+1}^2}{2} + \left(1 - \frac{U_{N+1}^2}{gh_0}\right) gh'_1$$

As the Froude number U_{N+1}^2/gh_0 is very small, the expression simplifies to:

$$\frac{U_{sea}^2}{2} = \frac{U_{N+1}^2}{2} + gh'_1 \qquad [4.3]$$

By summing the energy balances for the N volumes V_i, we deduce:

$$\left\{1 - \frac{U_{N+1}^2}{gh_0}\right\} \frac{h'_1}{h_0} = \frac{c_p S}{20D} \frac{U_{N+1}^2}{gh_0} \left\{N - 3\frac{\sum_{i=1}^{N} h'_i}{h_0}\right\}$$

The two dominant terms, because the variations h'_i are small and the Froude number is low, are:

$$\frac{h'_1}{h_0} \cong \frac{c_p S}{20D} \frac{U_{N+1}^2}{gh_0} N \qquad [4.4]$$

The comparison of expressions [4.3] and [4.4] makes it possible to determine U_{N+1} and h'_1:

$$U_{N+1} = \frac{U_{sea}}{\sqrt{1 + \frac{Nc_p S}{10Dh_0}}}$$

$$h'_1 = \frac{U_{sea}^2}{2g} - \frac{U_{N+1}^2}{2g}$$

Equation [4.1] then makes it possible to calculate all values h'_i. Finally, equation [4.2] makes it possible to determine all values U'_i.

The numerical results for the marine current turbines farm are given in Table 4.4. Figure 4.15 also shows the variations in speed and water depth in the channel of the farm.

U_{sea}	N	c_p	D	S	h_0	U_{N+1}	$\wp_{turbine}$	Number of marine current turbines	$\wp_{farm}$
m/s	–	–	m	m^2	m	m/s	MW	–	MW
2	20	0.4	15	176.7	40	1.80	0.206	120	24.7

Table 4.4. *Speed of the current at the outlet of the marine current turbines farm and powers recovered by a turbine and by the farm estimated by considering the modification of the current in the farm*

The most important result of this exercise is the reduction of the current speed in the channel of the marine current turbines farm. There is an *effect to avoid passing through the farm*. Due to the head loss of fluid caused by the recovery of energy by the marine current turbines, a fraction of the flow bypasses the obstacle formed by farm. The phenomenon is analogous to the reflex of a car driver who changes lanes when he notices an accumulation of cars in front of him. Changing lanes is effective as long as other drivers do not take the same initiative; the flows adjust to establish a new balance between the traffic lanes.

Quantitatively, the reduction in current is limited since the current in the farm is reduced to 1.8 m/s while the speed of the sea current is 2 m/s. This is a reduction of 10% in the speed of the current when the flows pass through 20 marine current turbines.

However, the effect on the recovered power is much greater in relative value. The table also gives the power recovered by a marine current turbine and the total power recovered by the farm. By comparing with the powers estimated in question 1, we see that the power recovered is nearly 30% lower than that estimated if the reduction in current is not taken into account. To understand this, just remember that the power recovered by a turbine varies with the cube of the speed:

$$\wp_{turbine} = c_p \rho \frac{U^3 S}{2}$$

Figure 4.16 shows that the water depth and speed vary very little inside the channel containing the turbines. The slowing of the fluid occurs upstream of the marine current turbine farm. A (counterintuitive) result to highlight is that the fluid accelerates inside the channel while the water depth decreases. This is a remarkable and classic result for free surface flow. For a fluvial flow, the height of water decreases together with the reduction of the head, while the velocity increases. The opposite result is observed when the flow is in a torrential regime.

7) Let us return to the kinetic energy balance on a volume V_i while keeping the terms relating to turbulence:

$$Q\left\{\rho g(h_i - h_{i+1}) + \frac{1}{2}\rho(U_i^2 - U_{i+1}^2) + \frac{1}{2}\rho\left(u_{rms,i}^2 - u_{rms,i+1}^2\right)\right\} = c_p \frac{1}{2}\rho U_i^3 S + \iiint_{V_i} \rho\varepsilon dv$$

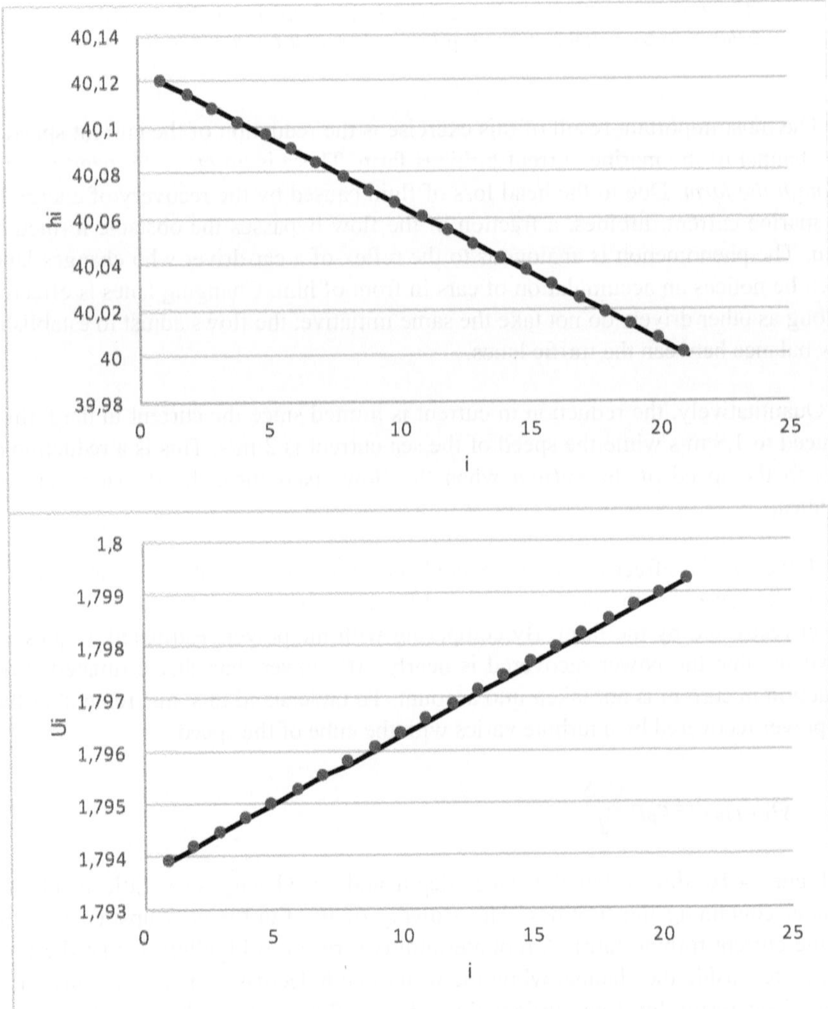

Figure 4.16. *Variation of water level and speed in a marine current turbines farm within an infinite marine space*

By summing over all volumes V_i, we deduce:

$$Q\left\{\rho g(h_1 - h_0) + \frac{1}{2}\rho(U_1^2 - U_{N+1}^2) + \frac{1}{2}\rho\left(u_{rms,1}^2 - u_{rms,N+1}^2\right)\right\}$$
$$= C_p \frac{1}{2}\rho S \sum_{i=1}^{N} U_i^3 + \iiint_{\sum_{i=1}^{N} V_i} \rho\varepsilon dv$$

A complete resolution of the problem requires having a turbulence model which determines how the rate of dissipation ε of kinetic energy is distributed in space. The terms to the right of the equality are both positive. Consequently, the kinetic energy dissipated will be added to the energy captured by the marine current turbines. Taking turbulence into account will therefore contribute to further reducing the current in the marine current turbines channel compared to what was calculated in question 6.

By summarizing all volumes, we notice that the turbulent kinetic energies compensate each other. On the entire volume of the marine current turbine farm, only the flow of turbulent kinetic energy $u_{rms,1}^2$ at the input and the flow of turbulent kinetic energy $u_{rms,N+1}^2$ at the output remain. It can be assumed that the turbulent kinetic energy at the entrance is zero and that which remains at the exit is that which has not been dissipated. Turbulent kinetic energy is produced during the energy recovery process by the turbines and in the wakes behind the machines. It results from the fact that the energy conversion process by the marine current turbines is not perfect. Part of the kinetic energy is lost in hydrodynamic head losses. The balance equation indicates that this loss of energy, caused by the non-ideal operation of the machines, is taken from the mean flow in the sea.

5

Hydroelectric Power

5.1. Hydroelectric power history and data

A hydroelectric plant is a facility that produces electricity using turbines that convert the potential energy of water contained in a basin at an altitude higher than the turbines. Hydroelectricity is a renewable energy production sector since basins, lakes and rivers are continuously and naturally supplied by the water cycle (evaporation and recovery in rainwater reservoirs in the watersheds).

Water mills used to grind grain are the ancestors of hydroelectricity. It has been known and used for over 2,000 years. Since the middle of the 19th century, with the start of industrial development, hydroelectricity developed into an industrial sector. In 1878, the term "white coal" was introduced in France to designate this energy resource. France played its part, notably by contributing to the design of high-efficiency turbines. The railway companies, which had significant energy needs, became directly involved in the construction of hydroelectric power stations, as evidenced today by some factory pediments.

There are two types of hydroelectric power plants: gravity power plants and pumped storage hydropower (PSH) plants.

The function of gravity plants is to recover using turbines the potential energy of water collected at altitude. They are classified into three categories: high head plants (more than 200 m), medium head plants (between 50 and 200 m) and low head plants (less than 50 m). This classification is schematically superimposed on the classification of turbines: Pelton turbines for high heads, Francis turbines for medium heads and Kaplan turbines for low heads (see section 2.3.1).

High and medium head gravity plants are always supplied by penstocks. Penstocks participate in energy conversion processes: they convert the potential energy of the fluid particles extracted from the upstream reservoirs into pressure energy at the inlet of the turbine (Figure 2.11). The speed of the flow in the penstock is limited, and consequently its kinetic energy, so that the dissipation of energy by the linear head losses in the penstock does not excessively reduce the energy recovered by the turbines.

Low-head gravity power plants operate on the course of a river or in the sea to harvest energy from the tide (Chapter 4). The pressure variation between the inlet and the outlet of the turbine is reduced. They only produce significant powers when the flow rate is high.

A complementary classification ranks gravity power plants according to the reservoir filling time, which is the time required to exhaust the reserve if the plant was operating at full power:

– run-of-river power plants: the reservoir is emptied in less than 2 h;

– medium reservoir power plants: the emptying of the reservoir lasts between 2 and 100 h;

– large reservoir power plants: the reservoir is emptied in more than 100 h.

The exercises proposed in this chapter concern high and medium head hydroelectric power plants. The theme of low-head power plants was covered in Chapter 4 devoted to tidal energy. In section 4.3, a Kaplan turbine sizing exercise was carried out.

PSH plants have a binary operation. Some of them were developed in France to meet the constraints of nuclear energy. As it is difficult to vary the production of nuclear power plants over demand, the idea is to use electricity to raise water using a pump and store it in retention basins at altitude, when electricity production exceeds demand. Conversely, when demand is high and the price of electricity increases, the hydroelectric turbines recover the potential energy of water stored at high altitude to produce electricity. The exercise proposed in section 5.3 studies the operation of the PSH of the Grand'Maison dam in the department of Isère, commissioned in 1987. This equipment is remarkable for its installed power (1,800 MW) and by its availability (it starts in a few minutes).

The PSHs were not all designed in conjunction with the French nuclear power program. Commissioned in 1954, the Pragnères power station, in the Hautes-Pyrénées, has an installed capacity of 185 MW. To exploit an extensive catchment area, but which did not allow the creation of a large capacity reservoir under the Vignemale mountain (for geological reasons), the hydroelectric power station was equipped with a pumping installation to transfer the water resource collected into the Ossoue reservoir to the Néouvielle lakes where the storage capacities are significant. It is a remarkable installation by its network (40 km of tunneling) and its operation.

The management of water resources stored in reservoirs is an essential concern for managing hydroelectric power plants. The resource is far from unlimited. In the exercise proposed in section 5.3, we quantify this aspect of the operation of the Grand'Maison power plant. The distribution of hydraulic capacities according to the type of power plant (Figure 5.1) shows that the various categories of power plants all contribute significantly to the production of hydroelectricity.

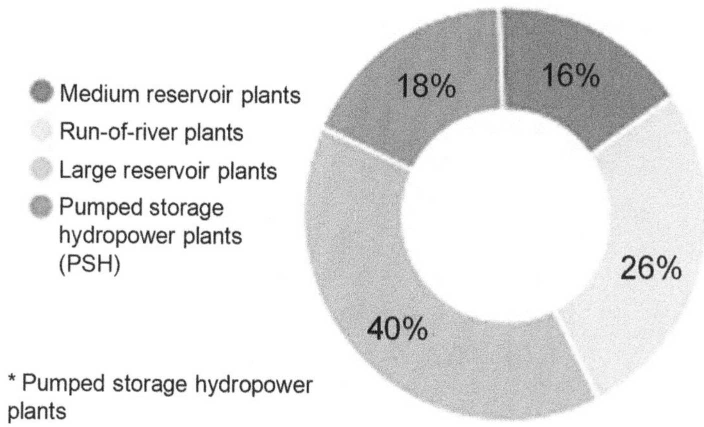

Figure 5.1. *Distribution of hydraulic capacities on the electricity transport network by type of power plant*[1]

The installed power of the French hydraulic fleet in 2021 amounts to 25,737 MW. Figure 5.2 shows the distribution by region as of June 30, 2021. The French territory is already very developed in terms of the exploitation of its hydroelectric resources. Undeveloped territories have poor development prospects due to the absence of mountain ranges or usable watercourses. The Renewable Energies Union mentions 37 MW of increase in connected power over 1 year as of June 30, 2021. Projects under development as this time represented 804 MW. Production over the year reached 59.9 TWh, corresponding to 12.8% of French electricity consumption. It is the second largest source of electricity production in France, behind nuclear power and ahead of wind power. The hydroelectricity load factor is 26.5%, which means that French hydroelectric production would be obtained if the plants operated at full power for just over a quarter of a year.

1 Source: "Panorama de l'électricité renouvelable au 30 juin 2021", publication of the Syndicat des Energies Renouvelables (SER).

Hydroelectricity is not always synonymous with gigantism. Small hydroelectricity refers to installations with an installed capacity of less than 10 MW. In France, it concerns around 2,270 power plants with 2,200 MW of installed power and 6 TWh of annual production on average[2].

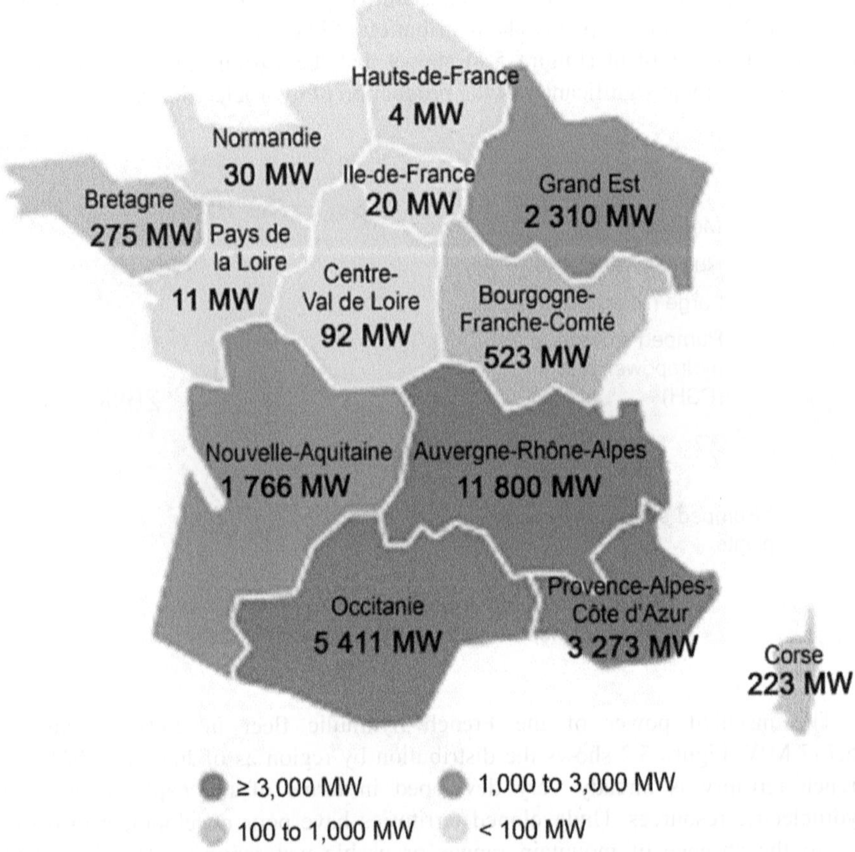

Figure 5.2. *Hydraulic power connected by region in France as of June 30, 2021*[3]

2 Available at: https://www.france-hydro-electricite.fr/lhydroelectricite-en-france/chiffres-clefs/.

3 Source: "Panorama de l'électricité renouvelable au 30 juin 2021", publication of the Syndicat des Energies Renouvelables (SER).

A global summary of hydroelectric production can be found by consulting Wikipedia[4]. Installed hydroelectric power worldwide reached 1,330 GW at the end of 2020 and hydroelectric production is estimated at 4,370 TWh. The share of hydroelectricity in world electricity production in 2019 is estimated at 15.6%. China is the leading producing country (31.0%), followed by Brazil (9.4%), Canada (8.8%), the United States (6.7%) and Russia (4.5%). France provides 1.5% of global hydroelectric production.

5.2. Exercise: maximum power provided by a hydroelectric power station

A penstock takes water from a mountain lake and brings it to the turbine of a hydroelectric power station located 400 m below. The pipe (made of cast iron) has a length of $L = 800$ m and a diameter $D = 1$ m. The turbine ejects the water into the open air through an orifice of the same diameter. Figure 5.3 schematizes the installation.

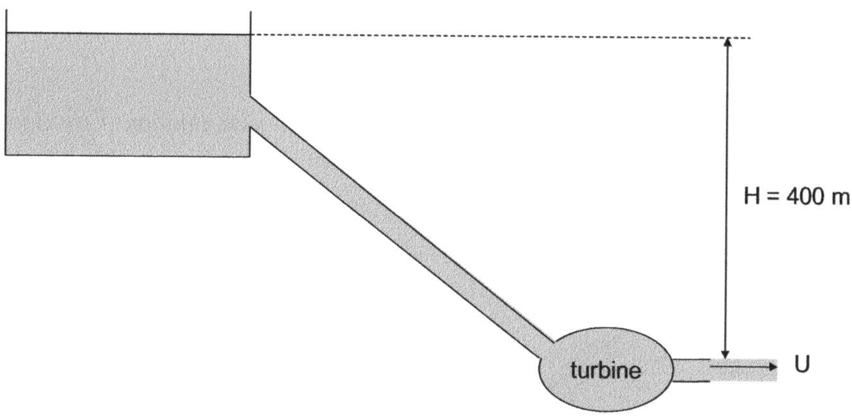

Figure 5.3. *Diagram of a hydroelectric power station*

1) Express the head variation $\Delta H_{turbine}$ between the inlet and outlet of the turbine as a function of the flow velocity U at the outlet of the turbine, taking into account the head losses in the hydraulic circuit.

4 Available at: https://fr.wikipedia.org/wiki/%C3%89nergie_hydro%C3%A9lectrique.

2) Express the hydraulic power of the turbine as a function of the flow speed U.

3) Determine the flow velocity U_{opt} in the pipe for the condition where the hydraulic power of the turbine is maximum. It will be assumed that the flow in the pipe is in the rough turbulent regime. Express the maximum theoretical hydraulic power of the plant and the $\Delta H_{turbine,opt}$ corresponding value. How much does the linear head loss reduce the collected head?

4) Carry out the numerical applications for the operation at maximum power of the hydroelectric plant: speed U_{opt}, flow rate Q_{opt}, head $\Delta H_{turbine,opt}$ and power $\wp_{max}$.

Assuming that the plant operates 5,000 h per year, calculate the energy supplied by the plant, the volume of water passing through the turbine and estimate the number of homes that could be supplied with electricity.

Are these operating conditions realistic?

5.2.1. *Solution*

1) The solution uses the generalized Bernoulli equation (section 1.6). The streamline followed by a fluid particle is drawn in Figure 5.4.

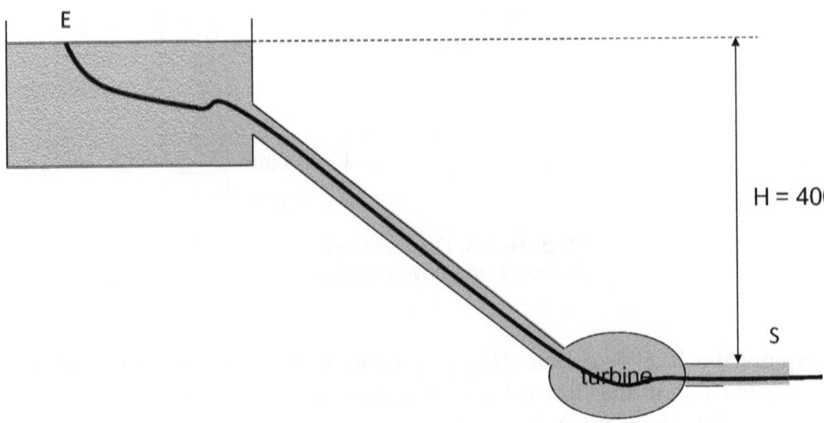

Figure 5.4. *Streamline (in gray) followed by a fluid particle in a hydroelectric plant*

The entry point E is chosen at the free surface of the upper basin. The fluid velocity at this point is zero (the basin is assumed to be very large). The head is there:

$$H_E = P_{atm} + \rho g H$$

for the altitude $z = 0$ taken at the level of the outlet of the turbine.

The point S is chosen at the outlet of the turbine. The water comes out of the turbine in the form of a jet in the open air. Its pressure is atmospheric pressure. The speed at the outlet of the turbine is equal to the flowing speed U in the penstock, by conservation of flow rate, since the diameter of the discharge pipe is identical to that of the penstock. The head in S is:

$$H_S = P_{atm} + \rho \frac{U^2}{2}$$

The variation in head of a fluid particle along the streamline between point E and point S is due to the linear head loss in the penstock and the head transferred to the turbine. Any singular pressure loss is neglected, particularly at the entrance to the penstock where the water is accelerated. We assume the entrance to the pipe is rounded so as to avoid flow separation.

Consequently, applying the generalized Bernoulli equation (section 1.6):

$$H_E - H_S = \rho g H - \rho \frac{U^2}{2} = \lambda \rho \frac{U^2}{2} \frac{L}{D} + \Delta H_{turbine}$$

and:

$$\Delta H_{turbine} = \rho g H - \rho \frac{U^2}{2}\left(1 + \lambda \frac{L}{D}\right)$$

2) The mechanical power recovered by the turbine is:

$$\wp = \Delta H_{turbine} Q = \left\{\rho g H - \rho \frac{U^2}{2}\left(1 + \lambda \frac{L}{D}\right)\right\} U \frac{\pi D^2}{4}$$

The power recovered is zero when $U = 0$, since the flow rate is zero, although the fluid particles give up all of the available head $\rho g H$.

Furthermore, when the speed is:

$$U_{max} = \sqrt{\frac{2gH}{1 + \lambda \frac{L}{D}}}$$

the available head is entirely dissipated in the pipe and $\Delta H_{turbine} = 0$. The recovered power is also 0 in this case.

The power recovered by the turbine is positive when the flow speed is in the interval:

$$0 < U < U_{max} = \sqrt{\frac{2gH}{1 + \lambda \frac{L}{D}}}$$

When the flow in the pipe is in the rough turbulent regime, the linear head loss coefficient λ does not depend on the velocity, but only on the relative roughness $k = e/D$ (see section 1.4). For a rusty cast iron pipe, the roughness of the pipe $e = 1$ mm is estimated. The roughness coefficient is equal to $k = 10^{-3}$ and the Colebrook diagram estimates the linear head loss coefficient $\lambda = 0.020$.

The power recovered is maximal for the delivery speed U such that:

$$\frac{d\wp}{dU} = 0$$

It is welcome to be able to consider that the flow in the pipe is in the rough turbulent regime[5], since the value of λ does not depend on U. In this case, the calculation of the derivative leads to:

$$\frac{d\wp}{dU} = \left\{\rho g H - 3\rho \frac{U^2}{2}\left(1 + \lambda \frac{L}{D}\right)\right\} \frac{\pi D^2}{4} = 0$$

Hence:

$$U_{opt} = \frac{U_{max}}{\sqrt{3}} = \sqrt{\frac{2gH}{3\left(1 + \lambda \frac{L}{D}\right)}}$$

5 The validity of the hypothesis is justified in question 4.

By replacing in the expression of $\Delta H_{turbine}$ (question 1):

$$\Delta H_{turbine,opt} = \frac{2}{3}\rho g H$$

The linear head loss consumes one-third of the available potential energy.

The maximum power is:

$$\wp_{max} = \frac{2}{3}\rho g H \sqrt{\frac{2gH}{3\left(1+\lambda\frac{L}{D}\right)}}\frac{\pi D^2}{4}$$

4) If the plant operates for an annual number of hours $n_{hours} = 5{,}000$ (which corresponds to a little more than half a year), the energy supplied (in Wh) is:

$$E_{annual} = \wp_{max} n_{hours}$$

and the volume of water passing through the turbine is:

$$Vol_{turb} = Q_{opt} \times 3{,}600 \times n_{hours}$$

Considering that the average annual electricity consumption of a household is worth 10,000 kWh, the number of homes supplied by this plant is deduced by dividing the annual energy supplied by the plant by 10,000 kWh.

The numerical applications are shown in Table 5.1.

λ	H	L	D	$\Delta H_{turbine,opt}$	U_{opt}	Q_{opt}
–	**m**	**M**	**m**	**Pa**	**m/s**	**m^3/s**
0.0195	400	800	1	2.62×10^{6}	12.55	9.86

$\wp_{max}$	n_{hours}	E_{annual}	Vol_{turb}	n_{hab}	Re
MW	**Hr**	**GWh**	**hm^3**	–	–
25.79	5,000	129	177	12,896	1.26×10^{7}

Table 5.1. *Operating conditions of a hydroelectric power plant with head height H = 400 m providing maximum power*

The volume of water passing through the turbine over one year is very large compared to the usual available resource. For comparison, the Grand'Maison reservoir (see section 5.3) has a volume of 140 hm^3 and the hydrological resource of the catchment area that feeds it is 105 hm^3. We always try to save the resource as much as possible, as we analyze in the exercise in section 5.3.

Hydroelectric power stations never work at the maximum power they could provide. As the linear head loss consumes one-third of the available head, the plant's energy efficiency is reduced to 67%. Plants usually operate with conditions that limit the head loss in the pipes in order to be efficient.

The operating conditions envisaged in this exercise are therefore not realistic. The exercise aimed to raise awareness of these issues and to work on the concepts of head and power.

The numerical applications carried out in this question make it possible to verify the hypothesis that the flow in the pipe is in the turbulent rough regime. The calculated Reynolds number for the velocity value (see Table 5.1) falls within the rough regime of the Colebrook diagram (section 1.4) for the roughness coefficient $k = 10^{-3}$.

5.3. Exercise: dimensioning of a PSH power plant

In this exercise, we design a PSH power station (PSH plant). The exercise is inspired by the Grand'Maison plant in Isère (France), near Grenoble. The data are taken from public documents on the Grand'Maison plant[6]. Photos of the Grand'Maison dam[7] and the power plant[8] make it possible to figure out the size of the equipment which has:

– an upstream reservoir with a capacity of 140 hm^3 closed by the Grand'Maison dam. It is supplied by an annual hydrological resource of 105 hm^3 (of which 87 hm^3 are collected between April and September) from a catchment area of 50 km^2;

– a downstream basin (Verney reservoir) with a capacity of 14.3 hm^3.

In the French energy context, when the production of electricity from nuclear sources is higher than demand, the water in the Verney basin can be pumped up to the Grand'Maison reservoir. In return, when demand requires it, the potential energy of water transferred from the Grand'Maison basin to the Verney reservoir is recovered by turbines to produce electricity. To carry out these operations, the power plant is equipped with four Pelton turbines and eight reversible turbine-pump groups.

The diagram of the power plant is reproduced in Figure 5.5.

6 Available at: http://energie.edf.com/fichiers/fckeditor/Commun/En_Direct_Centrales/Hydraulique/Centres/Les_Alpes/publications/documents/FicheamenagementGrandMaison.pdf (Jaumotte and Decock 1992).

7 Available at: http://www.hydrelect.info/articles.php?lng=fr&pg=122.

8 Available at: http://www.hydrelect.info/articles.php?lng=fr&pg=123.

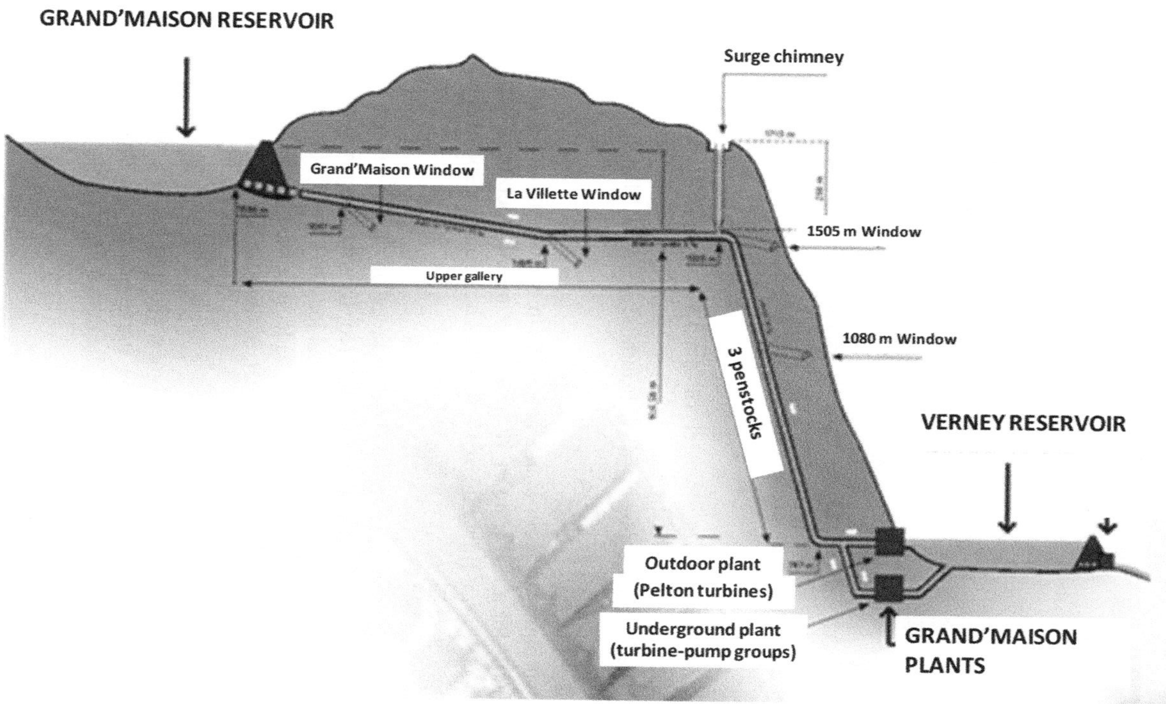

Figure 5.5A. *Diagram of the Grand'Maison hydroelectric power station (http://energie.edf.com/fichiers/fckeditor/Commun/En_Direct_Centrales/Hydraulique/Centres/Les_Alpes/publications/documents/FicheamenagementGrandMaison.pdf). For a color version of this figure, see www.iste.co.uk/mory/mechanical.zip*

Figure 5.5B. *Photograph of the Grand'Maison dam reservoir*[9]

The Pelton turbines have a wheel of diameter $D = 3.40$ m, equipped with 20 buckets[10]. Each is powered by five injectors and the rotation speed of the rotor is 428.6 rpm. The hydraulic power of each turbine is 157.1 MW for a flow rate of 19.5 m^3/s.

The pump-turbines are four-stage groups[11]. In turbine mode, each provides a power of 149 MW for a flow rate of 18 m^3/s. In pumping operation, each consumes a power of 163 MW for a flow rate of 17.8 m^3/s. Their rotation speed is 600 rpm and the rotor diameter is 2.25 m.

The difference in altitude between the surfaces of the lower and upper basins is $h =$ 930 m. The two basins are connected by two sections of pipes[12]. At altitude, a single, almost horizontal gallery of length $L_1 = 7{,}100$ m and diameter $D_1 = 7.1$ m constitutes the first section. It is made of concrete rock. It is extended by three parallel metal penstocks, almost vertical, of length $L_2 = 1{,}730$ m and diameter $D_2 = 3$ m.

In this exercise, the sizing of the turbines and pumps is carried out and the production and storage cycles that can be implemented on this installation are imagined. The results will be compared with the data quoted above.

9 Source: © EDF. All rights reserved.

10 See Jaumotte and Decock (1992).

11 Ibid.

12 Available at: http://energie.edf.com/fichiers/fckeditor/Commun/En_Direct_Centrales/Hydraulique/Centres/Les_Alpes/publications/documents/FicheamenagementGrandMaison.pdf.

5.3.1. *Part I – turbine study*

In this part, we size the plant's Pelton turbines. Since the power station provides a power of 1,800 MW at full speed, it is considered that each Pelton turbine provides an electrical power of 150 MW.

1) Write the expressions defining the overall efficiency of the plant and the efficiency of each turbine. Assuming that the overall efficiency of the plant is 0.85, determine the flow rate passing through each turbine. Also determine the flow velocities in the upper gallery and in the penstocks.

2) Calculate the linear head losses in the upper gallery as well as in the penstocks. Then calculate the efficiency of a Pelton turbine and the effective height of the installation.

3) Perform the dimensioning of two Pelton turbines for the operating conditions of the installation (flow rate and head) depending on whether the Pelton turbine has one injector or five injectors. The speed of the jets U_{jet} issuing from the injectors, the diameter d_{inj} of the jets and the diameter D_{rot} of the rotor will be determined. The form factor D_{rot}/d_{inj} will be taken equal to 15. Finally, calculate the rotation speed n and the specific speed n_s of the Pelton turbine. Note the effect of the number of injectors on the dimensioning of the turbine and compare the results of the dimensioning with the data on the Pelton turbines communicated at the beginning of the statement.

5.3.2. *Part II – study of pumping*

The technology of the eight reversible pump-turbine groups is confidential. It is not often that a manufacturer provides machines of this power. The data we were able to find[13] were summarized at the beginning of this exercise. Each pump–turbine group consists of four stages, which means that it is built with four identical machines placed in series. The dimensioning carried out in this part aims to study the characteristics of pumps capable of ensuring the same head and flow rate.

4) The Grand'Maison plant raises water from the Verney reservoir to the Grand'Maison reservoir using eight pump groups, each of which delivers a flow rate $Q_{pump} = 17.8\ \text{m}^3/\text{s}$. Determine the total upstream flow rate $Q_{storage}$, the velocities in the gallery and the penstocks. Recalculate the linear head losses and determine the head ΔH_{group} that the pump groups must supply to the fluid.

---

13 Available at: http://energie.edf.com/fichiers/fckeditor/Commun/En_Direct_Centrales/Hydraulique/Centres/Les_Alpes/publications/documents/FicheamenagementGrandMaison.pdf.

5) The rotors of the pump units rotate at speed $n = 600$ rpm. Determine the specific speed of the pump unit for raising the fluid with the flow rate $Q_{pump} = 17.8\ \mathrm{m^3/s}$, depending on whether the pump unit has one stage or four stages. Comment on the results.

6) The radius of the rotors of the four-stage pump units is $R_{rotor} = 1.125$ m. Calculate the head coefficient $\mu = \Delta H_{pump}/(\rho\omega^2 R_{rotor}^2)$ and the flow coefficient $\delta = Q_{pump}/(\omega R_{rotor}^3)$ of a pump in a group. Place the values on the diagram shown in Figure 2.7. Propose values of μ and δ for a single-stage pump unit that would perform the same function and deduce its rotational speed and its rotor diameter. Conclude with the benefit of using a four-stage pump unit.

5.3.3. *Part III* – *production plants and operation of the hydroelectric power station*

The hydraulic machines are placed near the Verney reservoir. The 4 Pelton turbines are in the outdoor plant, while the turbine-pump groups are in the underground plant below the water level of the Verney reservoir. The diagram of the production plant is shown in Figure 5.6, in which the diameters and lengths of the pipes are indicated. The altitude of the machines is also indicated.

7) Analyze the organization of the pipes and explain what the purpose of respecting the variations in pipe diameter is. By calculating the pressure at the inlet of the pump units, justify that the placement of the reversible turbine-pump units in the underground plant makes it possible to avoid cavitation in the pumping regime.

8) The power required by the pumps is such that they are started independently so as not to harm the network. The pumps are started using the electrical energy provided by the Pelton turbines. They are then powered by the electrical network when the synchronism speed is reached. Explain how this can be done. The installation is fitted with valves at the outlet of the Grand'Maison dam, which make it possible to stop the turbine flow. Another valve chamber at the junction between the upper gallery and the three descending penstocks also makes it possible to cut off the flows in the penstocks independently of each other. Give an estimate of the time needed to establish the pumping regime. What is the purpose of the surge chimney shown in Figure 5.5, at the junction between the gallery and the penstocks?

9) Considering the volumes of the reservoirs and the flow rates in the turbine and pumping regimes, estimate the emptying time of the Grand'Maison reservoir, the filling time of the Verney reservoir and the pumping emptying time of the Verney reservoir. Based on these estimates, propose the possible cycles at the Grand'Maison power plant.

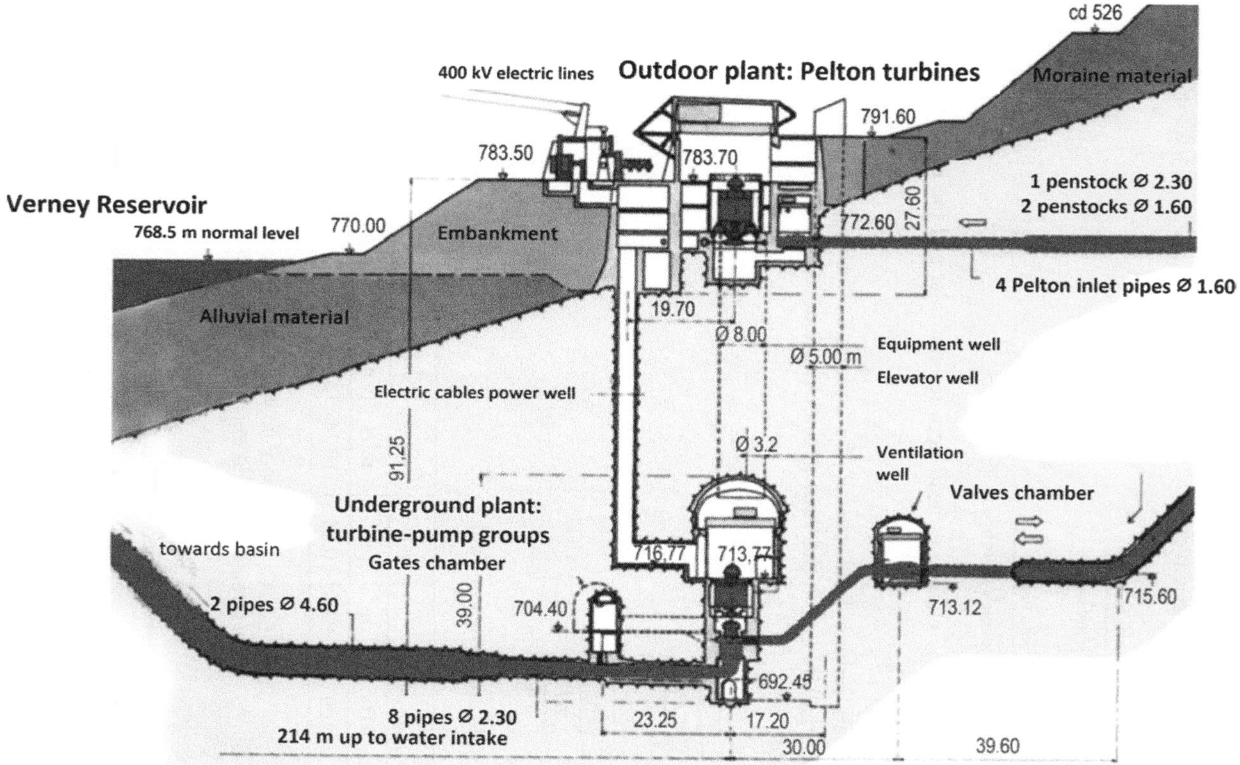

Figure 5.6. *Sectional view of the Grand'Maison plants (www.hydrelect.info/articles.php?lng=fr&pg=123). For a color version of this figure, see www.iste.co.uk/mory/mechanical.zip*

5.3.4. Solution

1) The overall efficiency of the installation and the efficiency of the Pelton turbine are expressed, respectively:

$$r_{global} = \frac{\wp_{elec,Pelton}}{\rho g h Q_{Pelton}}$$

$$r_{Pelton} = \frac{\wp_{elec,Pelton}}{\Delta H_{Pelton} Q_{Pelton}}$$

where ΔH_{Pelton} is the change in head between the inlet and the outlet of a Pelton turbine.

The overall turbine flow rate of the installation is $Q_{turb} = 12\, Q_{Pelton}$. The speeds in the gallery and in the three penstocks are deduced from the expression of the flow rate:

$$U_1 = \frac{Q_{turb}}{\frac{\pi D_1^2}{4}}$$

$$U_2 = \frac{Q_{turb}}{3\frac{\pi D_2^2}{4}}$$

The results are shown in Table 5.2.

$\wp_{elec,Pelton}$	h	r_{global}	Q_{Pelton}	Q_{turb}	D_1	U_1	D_2	U_2
MW	**m**	–	**m³/s**	**m³/s**	**m**	**m/s**	**m**	**m/s**
150	930	0.85	19.34	232.1	7.1	5.86	3.0	10.9

Table 5.2. *Calculation of flow rate in a Pelton turbine, total turbine flow rate and velocities in the pipes*

2) The upper gallery is made of rock-concrete and the penstocks are metallic. Assuming the latter are made of rusty cast iron, we can choose the same roughness e for the two pipes (section 1.4). Calculating the roughness coefficient $k = e/D$ and the Reynolds number $Re = UD/\nu$, the linear head loss coefficient is deduced from the Colebrook diagram (section 1.4) and the head loss is then calculated:

$$\Delta H_{lin} = \lambda \rho \frac{U^2}{2} \frac{L}{D}$$

Assuming that the outlet diameters of the Pelton turbines are chosen so that the speed at the outlet is equal to the speed in the penstocks, the head at the outlet of the Pelton turbine is $H_S = P_{atm} + \rho U_2^2/2$. As the head on the surface of the upper reservoir is $H_E = P_{atm} + \rho g h$, the change in head between the inlet and outlet of the Pelton turbine is:

$$\Delta H_{Pelton} = H_E - H_S - \Delta H_{lin,1} - \Delta H_{lin,2} = \rho g h - \rho \frac{U_2^2}{2} - \Delta H_{lin,1} - \Delta H_{lin,2}$$

(generalized Bernoulli equation, section 1.6). We deduce the efficiency of the turbine, according to the formula in question 1, which is the efficiency of the turbine independently of its insertion in the hydraulic circuit of the Grand'Maison plant:

$$r_{Pelton} = \frac{\wp_{elec,Pelton}}{\Delta H_{Pelton} Q_{Pelton}}$$

The effective height quantifies, in terms of water height, the head recovered by the turbine:

$$h_{eff} = \frac{\Delta H_{Pelton}}{\rho g}$$

The numerical values for each pipe are given in Table 5.3.

	e	D	L	U	k	Re
	mm	**m**	**m**	**m/s**	–	–
Gallery	1	7.1	7,100	5.9	1.4×10^{-4}	4.2×10^{7}
Penstock	1	3	1,730	10.9	3.3×10^{-4}	3.3×10^{7}

	λ	ΔH_{lin}	ΔH_{Pelton}	r_{Pelton}	h_{eff}
	–	**Pa**	**Pa**	–	**M**
Gallery	0.013	2.2×10^{5}	$8.32\ 10^{6}$	0.93	848
Penstock	0.013	5.2×10^{5}			

Table 5.3. *Calculations of the head losses in the pipes, the head recovered by the Pelton turbine, its efficiency and the effective height of the installation*

The numerical values show by consulting the Colebrook diagram that the flow in the gallery and the penstock is in a rough regime (the linear head loss coefficients do not depend on the Reynolds number, but only on the roughness coefficient). The effective height is 9% lower than the actual height h. The energy losses are limited, which is important to justify the operation of this plant as a PSH.

3) The sizing approach for a Pelton turbine is presented in section 2.3.2.

The speed of the jets leaving the injectors is:

$$U_{jet} = \sqrt{2gh_{eff}}$$

The diameter of the injectors is deduced from the flow rate, with n_{inj} the number of injectors of the turbine:

$$d_{inj} = \sqrt{\frac{4Q_{Pelton}}{n_{inj}U_{jet}\pi}}$$

The diameter of the rotor is then deduced from the form factor $D_{rot} = 15d_{inj}$.

The rotation speed of the wheel (in rpm) is then determined because the peripheral speed of the wheel must be half that of the jet to obtain the best hydraulic efficiency (equal to 1), that is:

$$n = \frac{30}{\pi}\frac{U_{jet}}{D_{rot}}$$

The specific speed of the Pelton turbine is given by the expression (section 2.1.7):

$$n_s = \frac{15}{\sqrt{\pi}}(2g)^{3/4}\sqrt{n_{inj}}\frac{d_{inj}}{D_{rot}}$$

In Table 5.4, we compare the sizing results depending on whether the Pelton turbine has one injector or five injectors. We verify that increasing the number of jets makes it possible to reduce the size of the turbine. Doubling the diameter of the rotor increases its volume and weight by a factor of 8; reducing the size of the machine is therefore a primary concern. The rotation speed increases in return, but it remains at an acceptable value for the five-injector turbine. Finally, let us add that a greater number of injectors allows better balancing of the forces applied to the wheel.

n_{inj}	h_{eff}	U_{jet}	d_{inj}	D_{rot}	n	n_s
–	**M**	**m/s**	**M**	**m**	**rpm**	–
1	848	129.0	0.437	6.55	188.0	5.3
5	848	129.0	0.195	2.93	420.3	11.8

Table 5.4. *Calculation of sizing of a Pelton turbine according to the number of injectors*

The calculated diameter of the five-injector turbine is a little below the real value of the Pelton turbines at the Grand'Maison plant (3.4 m) and the rotation speed is very close to the real value (428.6 rpm). For this installation, the simple tool for sizing a Pelton turbine still gives a satisfactory evaluation of the machine's characteristics.

4) The calculation is identical to that of questions 1 and 2, except that there are only eight pump groups, hence $Q_{storage} = 8Q_{pump}$. In the pumping regime, the velocities and head losses are lower in the gallery and the penstocks than they are in the turbine regime, although the linear head loss coefficients are identical because the flow regime is rough in the pipes.

The direction of flow is reversed in the pumping regime. The head on the surface of the Verney reservoir is $H_E = P_{atm}$, the same as it is $H_S = P_{atm} + \rho gh$ on the surface of the Grand'Maison basin. Between these two points, the fluid is supplied with head by the pump and loses it due to head losses in the gallery and in the penstocks. In addition, the widening at the entrance to the Grand'Maison reservoir produces a singular head loss, which dissipates the kinetic energy from the gallery. Hence:

$$\Delta H_{group} = H_S - H_E + \Delta H_{lin,1} + \Delta H_{lin,2} + \rho\frac{U_1^2}{2}$$

In turbine operation, head losses are energy that the turbines do not recover, while in pumping operation they are added to the potential energy that the pumps must supply.

Numerical values are shown in Table 5.5.

	e	D	L	U	k	Re	λ	ΔH_{lin}	ΔH_{group}
	mm	**M**	**m**	**m/s**	–	–	–	**Pa**	**Pa**
Gallery	1	7.1	7,100	3.6	1.4×10^{-4}	2.6×10^{7}	0.013	8.4×10^{4}	9.4×10^{6}
Penstock	1	3.0	1,730	6.7	3.3×10^{-4}	2.0×10^{7}	0.015	2.0×10^{5}	

Table 5.5. *Calculations of speeds and head losses in the pipes and of the head to be supplied by the pump group*

5) The specific speed is defined by:

$$n_s = n\frac{Q_{pump}^{1/2}}{\left(\Delta H_{pump}/\rho g\right)^{3/4}}$$

For a one-stage pump group, $\Delta H_{pump} = \Delta H_{group}$, whereas $\Delta H_{pump} = \Delta H_{group}/4$ when the group is made up of four pumps in series. The specific speed values are given in Table 5.6.

	Q_{pump}	ΔH_{group}	ΔH_{pump}	n	n_s
	m³/s	Pa	Pa	rpm	–
One-stage unit	17.8	9.41×10^6	9.41×10^6	600	14.7
Four-stage group	17.8	9.41×10^6	2.35×10^6	600	41.5

Table 5.6. *Calculation of the head to be supplied by the pump and its specific speed depending on whether the pump group has one stage or four stages*

Both pumps fall under centrifugal pump technology (Figure 2.6). The single-stage pump has a specific speed close to the lower limit of centrifugal pumps while the four-stage pump is at the upper limit.

6) The angular speed ω of rotation of the rotor is linked to its speed of rotation n (in rpm) by the relationship $\omega = 2\pi n/60$.

The head coefficient μ and the flow coefficient δ of a pump in a group relate, respectively, the head supplied by the pump and its flow rate to the angular speed of rotation and to the radius of the rotor (section 2.1.6):

$$\mu = \frac{\Delta H_{pump}}{\rho\omega^2 R_{rotor}^2}$$

$$\delta = \frac{Q_{pump}}{\rho\omega R_{rotor}^3}$$

These expressions can be inverted to determine the rotational speed of a pump and the rotor diameter, knowing the head coefficient μ and the flow coefficient δ of this pump:

$$R_{rotor} = \left(\frac{Q_{pump}}{\delta}\right)^{1/2} \left(\frac{\rho\mu}{\Delta H_{nom}}\right)^{1/4}$$

$$n = \frac{60\omega}{2\pi} = \frac{60}{2\pi}\left(\frac{\Delta H_{pump}}{\rho\mu}\right)^{3/4} \left(\frac{\delta}{Q_{pump}}\right)^{1/2}$$

	n	R_{rotor}	Q_{pump}	ΔH_{pump}	ω	μ	δ
	Rpm	**m**	**m³/s**	**Pa**	**rad/s**	–	–
Four-stage group	600	1.13	17.8	2.35×10^6	62.8	0.47	0.20
One-stage unit	1,117	1.15	17.8	9.41×10^6	116.9	0.52	0.10
One-stage unit	586	2.13	17.8	9.41×10^6	61.4	0.55	0.03

Table 5.7. *Values of the radius of the pump rotor and its rotation speed corresponding to the values of the head and flow coefficients*

The values of the head coefficient and the flow coefficient are given in Table 5.7 for the rotor radius and the rotation speed given in the documentation[14] for the four-stage pump group at the Grand'Maison plant. The diagram proposed by Rey et al. (2012; Figure 2.6) confirms the possibility of finding a pump of this type near the upper limit (in terms of flow coefficient) for centrifugal pumps.

In centrifugal pump technology, the diagram of Rey et al. (2012) indicates that the head coefficient can only be marginally increased, but the flow coefficient can be significantly decreased. For a single-stage pump, two pairs of values are proposed (μ, δ) in the table, chosen between the two dotted curves on the diagram of Rey et al. A single-stage unit with the same rotor diameter as the four-stage unit should spin nearly twice as fast. When keeping the same rotational speed, then the rotor diameter must be doubled on a single-stage unit compared to the rotor diameter of a pump in a four-stage unit.

Doubling the rotor diameter of a machine amounts to multiplying its volume by 8. We can therefore consider that using a four-stage group rotating at the same speed as a single-stage group allows the total volume of the machine to be divided by 2.

7) The pipeline network of the Grand'Maison plant is complex. We will return in question 8 to the subdivision of the upper gallery into three penstocks. At the bottom, the flow is divided into two branches, one of which leads to the outdoor plant and the other to the underground plant. Ultimate derivations lead to each machine. The pipe diameters are chosen to limit the delivery speed (and thus limit the linear head losses) and avoid its variations. A sudden decrease in the flow speed risks generating a singular head loss.

Water is pumped from the Verney basin by two diameter pipes $D_{coll} = 4.6$ m, which are divided into eight diameter pipes $D_{suc} = 2.3$ m supplying each turbine group. The diameters of the pipes are also chosen so as to avoid

14 See Jaumotte and Decock (1992).

variations in flow velocity. The flow velocity at the inlet of the pumps and in the collection pipe is, in pumping mode, $U_{suc} = Q_{pump}/(\pi D_{suc}^2/4)$. For a value of the estimated linear head loss coefficient $\lambda = 0.014$, the linear head loss is low for a collection pipe of length about 200 m. In pumping regime, the pressure at the inlet of the pump is, using the generalized Bernoulli equation (section 1.6):

$$P_{suc} = P_{atm} - \rho g Z - \rho \frac{U_{suc}^2}{2}\left(1 + \lambda \frac{L}{D_{coll}}\right)$$

The depth of the pumps-group ($Z \cong -60$ m) helps protect against the risk of cavitation at the pump inlet. The numerical values are given in Table 5.8.

Q_{pump}	D_{suc}	U_{suc}	D_{coll}	λ	L	Z	P_{suc}
m³/s	**m**	**m/s**	**m**	**–**	**m**	**M**	**Pa**
17.8	2.3	4.3	4.3	0.014	200	– 60	6.7×10^5

Table 5.8. *Calculation of pressure at the pump inlet*

8) The start-up of the reversible groups in a pumping operation is done using the electrical energy delivered by the Pelton turbine alternators. This presupposes simultaneous operation of the Pelton turbines (and therefore a downward flow of water) while an upward flow from the pumps begins. The usefulness of having three parallel penstocks appears in this process.

By closing the gates at the top of two of the penstocks, turbine flow can be maintained in the upper gallery and in a penstock leading to the outdoor plant to feed the Pelton turbines. The electricity produced by their alternators is used to start the eight reversible groups of the underground plant in pumping mode. Simultaneously with the rise of water in the other two penstocks, the pumping power gradually increases.

When the rising water reaches the top of the two penstocks, it continues to flow in the gallery, provided that the valve at the outlet of the Grand'Maison dam has been closed beforehand and that the water descending from the gallery has been evacuated in the descending penstock before the arrival of the ascending flow. The Pelton turbines can then be shut down and the pumping flow then distributed over the three penstocks, reducing speed and linear head losses.

The time taken by the fluid to move up the penstock is $T_{start} = L_2/U_{2,start}$, where $U_{2,start}$ is 1.5 times the speed in the penstock calculated in question 6, since the pumping flow passes through two penstocks while in the final regime it uses the three pipes.

L_2	U_2	U_{start}	T_{start}
M	**m/s**	**m/s**	**s**
1,730	6.7	10.1	171.8

Table 5.9. *Velocities in the vertical pipe when starting the pump*

It is indicated in the documentation a duration of 10 minutes to reach the steady state.

This calculation shows the importance of the valves for the operation of the plant. We also understand the usefulness of the surge chimney. By letting the fluid rise in the vertical pipe at the junction of the gallery and the penstocks near the valves obstructing the upper ends of the penstocks, it protects against water hammer and excess pressure produced by the closing of the valves.

9) The turbine and pumping flow rates of the Grand'Maison plant are considerable in view of the volumes available.

The duration for emptying the reservoir of the Grand'Maison dam, if it is full, is:

$$T_{empty\ GM} = \frac{Vol_{GM}}{Q_{turb}}$$

The duration of filling the Verney reservoir, if it is initially empty, is:

$$T_{fill\ V} = \frac{Vol_V}{Q_{turb}}$$

When the Verney reservoir is full, the flow must evacuate toward the river and the resource is lost for later storage.

The duration of pumping of the Verney reservoir, if it is initially full, is:

$$T_{empty\ V} = \frac{Vol_V}{Q_{storage}}$$

The results are shown in Table 5.10.

Vol_{GM}	Vol_V	Q_{turb}	$Q_{storage}$	$T_{empty\ GM}$	$T_{fill\ V}$	$T_{empty\ V}$
m ^{3}	m ^{3}	m ^{3}/s	m ^{3}/s	s	s	s
1.4×10^8	1.43×10^7	232.1	142.4	6.03×10^5	6.16×10^4	1.0×10^5
–	–	–	–	6 d 23 h 32 min	5 h 07 min	1 d 3 h 54 min

Table 5.10. *Duration of emptying of the Grand'Maison reservoir and filling of the Verney basin in turbine regime and duration of emptying of the Verney reservoir in pumping regime*

The resource must be managed economically. The total emptying of the Grand'Maison reservoir takes place in 7 days. The annual hydrological resource is estimated at 105 hm^3, lower than the volume of the reservoir. These figures show how considerable the power installed at the Grand'Maison plant is compared to the resource. Its operation only makes sense when it is operated in a PSH, with turbine and pumping cycles using the available electricity. Otherwise, the full power operation time would be less than 7 days per year.

A turbine or pumping operation lasts at most 1 day if the hydraulic resource is to be conserved.

6

Osmotic Energy

6.1. The phenomenon of osmosis

The phenomenon of osmosis is demonstrated by carrying out the experiment shown schematically in Figure 6.1. We consider a jar separated into two compartments by a dense porous wall. The membrane is semi-permeable: it allows the passage of water but it retains very small particles, such as ions.

The same volume of liquid is initially introduced (Figure 6.1(a)) into the two compartments. The right compartment contains pure water, while the left compartment contains salt water with a concentration of ions C_0. Salt water and pure water, which are in contact through the membrane, are not in chemical equilibrium. As the salt ions cannot cross the membrane to equalize the salt concentration on both sides, it is the water which crosses the membrane from the right compartment to the left compartment to reduce the difference in concentration by diluting the saline solution. This flow of pure water is the osmosis flow.

The osmosis flow then raises the water level in the left compartment while the water level falls in the right compartment. The speeds are very low and the pressure is hydrostatic in both compartments. The difference in piezometric pressure between the compartment containing the salt water and the compartment containing the pure water is $\Delta P = \rho g H$, where H is the difference in level between the two compartments. This pressure difference in return generates a filtration flow, called reverse osmosis flow, from the salt water compartment to the pure water compartment. By virtue of Darcy's law, the flow rate velocity increases linearly with H. A balanced situation is reached for the height $H = H_{osm}$ for which the osmotic flow rate is equal to the reverse osmotic flow rate (Figure 6.1(b)).

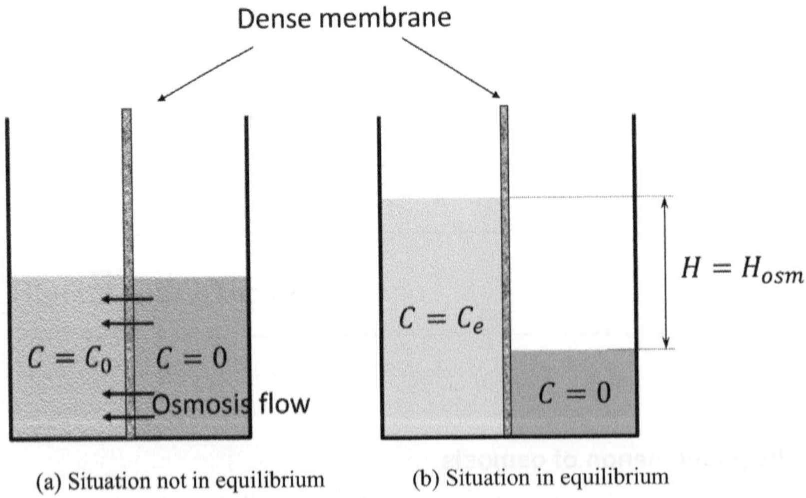

Figure 6.1. *Principle of an experiment highlighting the phenomenon of osmosis*

The experiment in Figure 6.1 is not simple to carry out in practice because the height of the column should be measured in hundreds of meters. The height H_{osm} measured in the experiment makes it possible to calculate the osmotic pressure ($P_{osm} = \rho g H_{osm}$), which is also calculated with the van't Hoff equation for osmotic pressure:

$$P_{osm} = C_e RT$$

This equation is formally identical to the ideal gas law. T denotes the absolute temperature and $R = 8.32$ J/mol/K is the ideal gas constant. C_e is the concentration (in mol/m^3) of particles. Salts dissolve in water as ions. As an example, each molecule of sodium chloride NaCl dissolves into a Na^+ ion and a Cl^- ion. The particle concentration is therefore twice the molar concentration of NaCl.

Osmosis is a natural phenomenon. It plays its part in governing our lives since osmosis regulates the exchange of liquid between the inside of our cells and the outside. A more directly visible example is that of a plant. When a plant is watered with pure water, the water on the surface of the leaves is absorbed into the plant cells through a process of osmosis. But if the plant is watered with sea water, it can dry out if the salt concentration in the sprayed drops is higher than the salt concentration in the cells. Let us cite a final example from the kitchen. Salt is used to dry a ham or a duck breast. It causes an osmotic flow from the less salty interior of the meat to the saline environment in which the piece of meat is placed.

Reverse osmosis is used to desalinate seawater and produce drinking water. A flow in the opposite direction to that of osmosis is caused by applying a pressure difference greater than the osmotic pressure between the salty side and the non-salty side of a dense membrane.

When the pressure difference between the salty and unsalted side of a dense membrane is less than the osmotic pressure, the osmosis flow dominates. The exercise proposed in section 6.2 sizes such an installation to produce electricity. For this exercise, we use a module used for seawater desalination, which is operated under conditions where the osmosis flow dominates. To assess the relevance of the energy production process, it is interesting to compare its performance to that of the same module used to desalinate seawater.

6.2. Exercise: sizing an electricity production system using osmotic energy

The process is schematized in Figure 6.2. Unsalted water brought by a river is in contact through a dense membrane with sea salty water contained in a volume called a reactor. This membrane is selective; it lets water molecules pass but retains salt ions. Osmotic flow occurs when the piezometric pressure P_{int} inside the reactor is less than $P_{riv} + P_{osm,int}$. P_{riv} is the piezometric pressure in the river and $P_{osm,int}$ the osmotic pressure in the reactor. The water level then rises in the reactor to the threshold whose height is H above the surface of the river and the sea. This threshold is set such that $H < H_{osm,int} = P_{osm,int}/\rho g$. The water then returns to the sea passing through a turbine which converts the pressure energy in the reactor to produce electricity.

In the real process, the pressure level in the reactor is not controlled by a physical threshold as shown schematically in Figure 6.2, but by the characteristic curve $\Delta H_{turb}(Q_T)$ of the turbine. By representing on the diagram the head in the reactor by the height H of a threshold, we schematize that this height H is a process adjustment parameter.

In this exercise, the power delivered by the process is determined and its value is optimized. The only head variations taken into account are the head variation of the fluid particles passing through the membrane, the head variation across the pump and the head variation across the turbine. The efficiencies of the turbine and the pump are assumed to be equal to 1. The linear head losses in the pipes are also neglected.

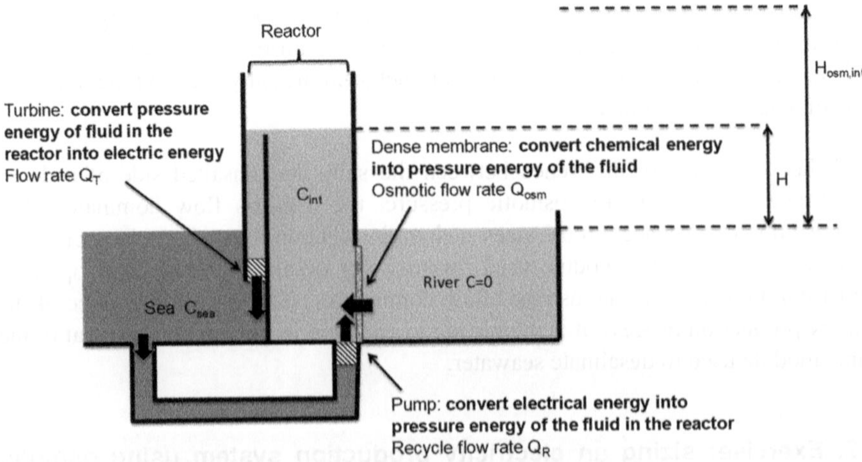

Figure 6.2. *Diagram of a process for producing electricity from osmotic energy*

1) Concentration of NaCl salt in the sea is taken equal to 35 g/L. The molar masses of sodium and chlorine are 23 g/mol and 35 g/mol, respectively. Calculate the osmotic pressure $P_{osm,sea}$ in the sea at temperature $T = 300$ K. The dissociation of a NaCl molecule into two ions Na^+ and Cl^- must be taken into account.

2) Mixers ensure the homogeneity of the salt concentration C_{int} in the reactor. Explain why a recycling flow Q_R from the sea is necessary to ensure permanent electricity production. Write the mass balance over the volume of the reactor and deduce a relationship between the flow rate Q_{osm} through the membrane, the flow rate Q_T through the turbine and the recycle flow rate Q_R.

3) Express the NaCl salt material balance on the volume of the reactor and deduce a relationship linking the salt concentration C_{int} in the reactor to the salt concentration C_{sea} in the sea. Express the osmotic pressure $P_{osm,int}$ in the reactor as a function of the osmotic pressure $P_{osm,sea}$.

4) Determine the fluid head H_{riv} in the river, the fluid head H_{sea} in the sea and the fluid head H_{int} in the reactor. Then express the change in head ΔH_{osm} of a particle passing through the membrane, the change in head ΔH_{pump} of a fluid particle passing from the sea to the reactor through the recycling pump and the change in head $\Delta H_{turbine}$ of a fluid particle passing from the reactor to the sea through the turbine. Finally, express the powers associated with the pump, the turbine and the osmotic flow.

5) The dense membrane is not a flat membrane as shown schematically in Figure 6.2. We use a module described by Remigy and Desclaux (2007), which is used for the desalination of sea water[1]. The membranes are hollow tubes (Figure 6.3) with an interior radius $R_i = 20$ µm and an exterior radius $R_e = 40$ µm. Their thickness is therefore $e = 20$ µm. A multitude of parallel tubes are manufactured in a module (Figure 6.4). The salt water passes inside the tubes. This corresponds to the interior of the reactor in Figure 6.2 and the head is H_{int}. River water enters at the top and is distributed among the fibers. The head is H_{riv}. Due to the difference in head between the water in the reactor and the water in the river, the osmosis flow passes radially through the fibers as indicated by the arrows in Figure 6.3. The filtration device is a tangential filtration process. The salty water flow flows inside the hollow fibers and comes out the other side. This limits the deposit of salt and the clogging of the membrane.

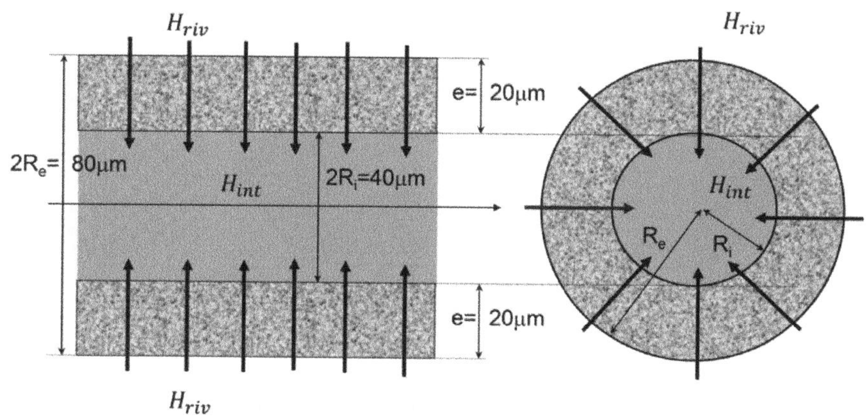

Figure 6.3. *Geometry and dimensions of hollow filtration fibers*

The circular cylindrical geometry of the fibers requires adapting Darcy's law:

$$\mu \frac{Q_{osm}}{S} = k \frac{\Delta P - P_{osm}}{e}$$

1 When the module is used to desalinate sea water, the salt water circulating inside the fibers of the module is sea water. By imposing $H_{int} - H_{riv} > P_{osm,sea}$, a reverse osmosis flow is produced. The modifications to be made are to reverse the direction of flow through the membrane in Figure 6.3 and to indicate a flow of unsalted water exiting from the top in Figure 6.4.

so as to ensure flow conservation. The osmotic flow Q_{osm} through the hollow fibers is deduced from:

$$\mu \frac{Q_{osm}}{2\pi L R_e}\left(\frac{R_e}{e} Ln\left(\frac{R_e}{R_i}\right)\right) = k \frac{P_{osm,int} - \Delta H}{e}$$

where $\Delta H = H_{int} - H_{riv}$ is the change in head across the membrane. The permeability of the material constituting the fibers is $k = 10^{-21}$ m^2, $e = 20$ µm the thickness, and $S = 2\pi L R_e = 15{,}000$ m^2 is considered as the filtration surface. The dynamic viscosity of water is $\mu = 10^{-3}$Pa. s. Express the power recovered by the process in a form where H is the only unknown.

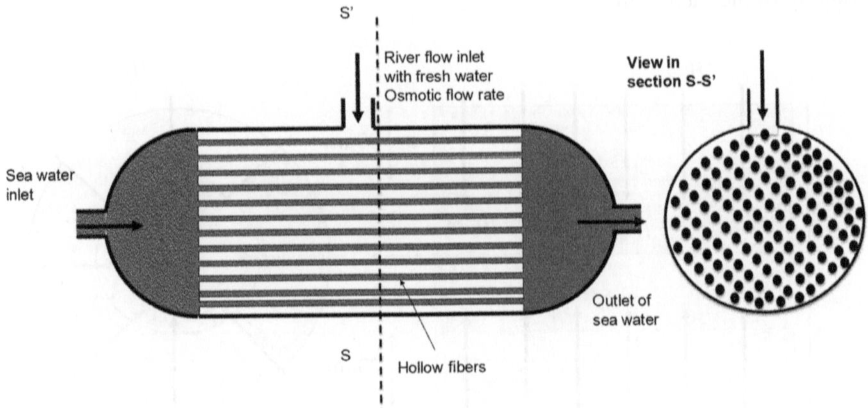

Figure 6.4. *Hollow fiber module*

What is the point of having such a low fiber thickness e?

6) The recycling flow rate Q_R is chosen so that $C_{int} = 3C_{sea}/4$. Determine the value H for which the energy produced by the process is maximum. Then determine the flow rates Q_{osm}, Q_T and Q_R, and the power supplied by the process.

7) The module is very compact. Remigy and Desclaux (2007) indicate that a module with a volume of 1 m^3 can contain fibers reaching a total filtration surface of 15,000 m^2. For the configuration described in Figures 6.3 and 6.4, calculate the total length L of the fibers for a total filtration surface area $S = 15{,}000$ m^2. Estimate the volume filled by the fibers. Comment on the values obtained. What is the point of having such small fiber radii?

8) The large pressure difference between the inside and outside of the fibers questions the mechanical resistance of such fine fibers. Figure 6.5 schematizes the mechanical equilibrium of a fiber element of length L (in the direction perpendicular to the figure) for an arc element $2d\theta$. The fiber does not break if the pressure forces applied to the inner and outer faces of the fiber are balanced by the tension force $\vec{T}$ that stretches the fiber. By writing the balance of forces on the fiber element, calculate the stretching stress σ (defined by $T = \sigma Le$) as a function of piezometric pressures P_{int} and P_{riv}. Hollow fibers are made from cellulose acetate or polyamide. For these materials, after having researched the values of their tensile strength (limit stress which causes them to break), check their mechanical resistance for the operating conditions of the module and explain why the fiber does not break.

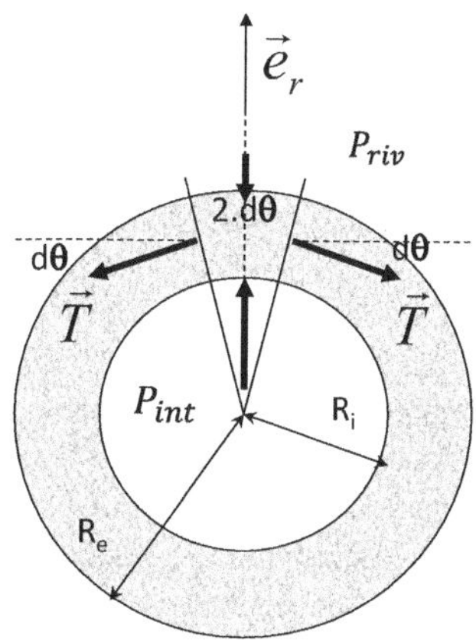

Figure 6.5. *Mechanical equilibrium of a fiber element of arc length $2d\theta$, thickness e and length L in the direction perpendicular to the figure*

9) Draw your conclusions on this process of producing electricity from osmotic energy by analyzing in particular:

– the technological level required;

– the power produced.

To provide material for comparison, the same module can supply 700 houses with drinking water (assuming a consumption of 500 L per day per house) with a cost of 2 kWh per m^3 of drinking water produced[2].

2 The sizing of the seawater desalination process by reverse osmosis using the same module can be carried out by solving the exercise proposed in chapter 14 of Mory (2011).

6.2.1. *Solution*

1) The salt concentration expressed in g/L must be converted into a concentration in mol/m^3:

$$C_{sea} = 2\,\frac{C_{sea,g/L}.\,\mathrm{g.\,mol.\,1{,}000L}}{(35+23).\,\mathrm{L.\,g.\,m^3}} = 1{,}207\,\frac{\mathrm{mol}}{\mathrm{m^3}}$$

The factor 2 in front of the quotient takes into account the dissociation: a molecule of NaCl produces two ions or two particles.

The osmotic pressure in the sea is:

$$P_{osm,sea} = C_{sea}RT = \rho g H_{osm,sea} = 3\,\times\,10^6\mathrm{Pa} = 30\ \mathrm{bar}$$

This is a significant pressure level. Whether to desalinate seawater or to produce electricity, as envisaged in this exercise, the membranes are subjected to high pressure levels.

2) Let us imagine on the diagram in Figure 6.2 that the system operates without recycling sea water to the reactor and that we start with a reactor containing sea water up to the same height as at sea. The osmotic flow causes the water to rise to the threshold. When it is reached, the water contained in the reactor flows toward the sea passing through the turbine. As an osmotic flow rate is continuously introduced, the salt concentration in the reactor decreases over time. Osmotic pressure also drops and osmotic flow through the membrane decreases. To maintain electricity production, it is therefore necessary to maintain the salt concentration in the reactor. This is what the recycle flow Q_R is for.

Mass balance on the volume of water V_{react} contained in the reactor (neglecting the variation in density with the salt content) in steady state:

$$\frac{d(\rho V_{react})}{dt} = \rho Q_{osm} + \rho Q_R - \rho Q_T = 0$$

Hence, the relationship between the volume flow rates:

$$Q_{osm} + Q_R = Q_T$$

3) Material balance on the quantity of salt $C_{int}V_{react}$ contained in the reactor in steady state:

$$\frac{d(C_{int}V_{react})}{dt} = 0 \times Q_{osm} + C_{sea}Q_R - C_{int}Q_T = 0$$

Hence, the relationship:

$$\frac{C_{int}}{C_{sea}} = \frac{Q_R}{Q_T}$$

van't Hoff's law being a linear relationship, the osmotic pressures are related to the concentrations, hence:

$$P_{osm,int} = \frac{C_{int}}{C_{sea}} P_{osm,sea}$$

4) Except in the vicinity of the pump and the turbine, the speeds in the sea, river and reactor are low. The pressure distribution is hydrostatic, and the head is almost equal to the piezometric pressure. The head is homogeneous in the three volumes. It is estimated in each volume by calculating it at the free surface where the pressure is atmospheric pressure.

Taking $z = 0$ at the surface of the river (which is at the same level as the sea), we deduce:

$$H_{riv} = H_{sea} = P_{atm}$$

$$H_{int} = P_{atm} + \rho g H$$

The variation in head ΔH_{osm} of the fluid particles crossing the membrane coming from the river is:

$$\Delta H_{osm} = H_{int} - H_{riv} = \rho g H$$

The variation in head ΔH_{pump} of the fluid particles introduced by the pump from the sea to the reactor is:

$$\Delta H_{pump} = H_{int} - H_{sea} = \rho g H$$

The variation in head $\Delta H_{turbine}$ of the fluid particles returning from the reactor to the sea through the turbine is:

$$\Delta H_{turbine} = H_{int} - H_{sea} = \rho g H$$

It is strictly negative; its absolute value is given above.

The powers associated with the pump, the turbine and the osmotic flow are:

$$\wp_{osm} = \Delta H_{osm} Q_{osm} = \rho g H Q_{osm}$$

$$\wp_{pump} = \Delta H_{pump} Q_R = \rho g H Q_R$$

$$\wp_{turbine} = \Delta H_{turbine} Q_T = \rho g H Q_T$$

5) Some remarks on Darcy's law through a planar membrane and a membrane of circular geometry:

– the term in parentheses containing the logarithm is equal to $2Ln(2)$ for the circular geometry of hollow fibers (Figure 6.3) and has no dimension. Other terms can be put into direct correspondence with Darcy's law for a plane membrane;

– for circular geometry, we changed the sign of the right-hand side so that the flow rate is positive when it is an osmotic flow;

– the definition of the filtration surface $S = 2\pi L R_e$ is a choice. Any other circular cylindrical surface $2\pi L r$ could have been chosen, since the flow crosses all these surfaces for $R_i < r < R_e$;

– the permeability of the material $k = 10^{-21}\text{m}^2$ is very low for a dense membrane. It lets water molecules pass but retains salt ions. The size of its pores is very small, of the order of an angstrom. Permeability is often called intrinsic permeability to recall that it is a characteristic of the material alone (and not of the fluid passing through it). As its dimension is in m^2, the permeability only depends on geometric characteristics: the size of the pores d and the porosity of the material (dimensionless). k is therefore proportional to d^2. This explains the low value of k. It is hardly possible to increase the permeability of the membrane while preserving its selectivity.

Darcy's law allows us to express the osmotic flow rate:

$$Q_{osm} = S\frac{k}{\mu}\frac{P_{osm,int} - \rho g H}{e2Ln(2)}$$

We deduce from this formula that reducing the thickness e is the lever allowing the most effective increase in osmotic flow.

The power supplied by the process is the difference between the power recovered by the turbine (from which the electricity can be sold) and the power consumed by the pump (we must pay for the electricity to operate it). The osmotic power does not explicitly enter into this calculation since we neither consume nor produce electricity by crossing the membrane. Replacing the different powers by their values, we deduce:

$$\wp_{process} = \wp_{turbine} - \wp_{pump} = \rho g H Q_T - \rho g H Q_R = \rho g H Q_{osm} = \wp_{osm}$$

This result is obtained in the ideal hypothesis where the friction head losses and singular losses are negligible. This ideal case makes it possible to identify the role of the different converters: the membrane converts chemical energy into mechanical energy (high pressure), the turbine converts mechanical pressure energy into electricity and the pump consumes energy electricity to provide pressure energy to recycled seawater. These energy conversion processes constituting the process are indicated in bold in Figure 6.2.

By expressing Q_{osm} in the power expression, we finally obtain:

$$\wp_{process} = \frac{k}{\mu}\frac{S}{e2Ln(2)}(P_{osm,int} - \rho gH)\rho gH$$

Once the ratio of recycling and turbine flow rates have been fixed:

$$\frac{C_{int}}{C_{sea}} = \frac{Q_R}{Q_T}$$

$$P_{osm,int} = P_{osm,sea}\frac{C_{int}}{C_{sea}}$$

the only parameter left to choose is the height H.

6) For the desired operating conditions:

$$\frac{C_{int}}{C_{sea}} = \frac{3}{4} = \frac{Q_R}{Q_T}$$

$$P_{osm,int} = P_{osm,sea}\frac{C_{int}}{C_{sea}} = P_{osm,sea}\frac{3}{4}$$

$$\wp_{process} = \wp_{osm} = \frac{k}{\mu}\frac{S}{e2Ln(2)}(P_{osm,int} - \rho gH)\rho gH$$

Maximum power is recovered when:

$$\frac{d\wp_{osm}}{dH} = 0$$

that is to say when:

$$\rho gH = \frac{P_{osm,int}}{2}$$

Hence, the maximum power produced:

$$\wp_{process,max} = \frac{k}{\mu}\frac{S}{e2Ln(2)}\frac{P_{osm,int}^2}{4}$$

The osmotic flow rate is then:

$$Q_{osm} = S\frac{k}{\mu}\frac{P_{osm,int}}{4eLn(2)}$$

The flow rates are linked by the relationship:

$$Q_{osm} = Q_T - Q_R = Q_T\left(1 - \frac{Q_R}{Q_T}\right) = \frac{Q_T}{4}$$

The numerical results are given in Table 6.1.

k	e	μ	S	$P_{osm,sea}$	$\frac{C_{int}}{C_{sea}}$	$\wp_{osm}$	Q_{osm}	Q_R	Q_T
m^2	m	Pa.s	m^2	Pa	–	kW	m^3/day	m^3/day	m^3/day
10^{-21}	2×10^{-5}	10^{-3}	15,000	3×10^6	0.75	0.68	105	316	421

Table 6.1. *Numerical values of the operating parameters of the process of electricity production by conversion of osmotic energy*

7) The filtration surface $S = 2\pi L R_e$ allows the total length of fibers to be calculated:

$$L = \frac{S}{2\pi R_e}$$

The volume occupied by these fibers is then:

$$Vol_{fibres} = \pi R_e^2 L = \frac{S R_e}{2}$$

The numerical applications are given in Table 6.2.

The length of the fibers is gigantic (almost twice around the Earth), but the volume of the fibers fits into 1 m^3. We measure the technological prowess necessary to create such a module. The costs of the technology are undoubtedly essentially those of manufacturing the fibers and assembling them into a module.

S	R_e	L	Vol_{fibres}
m^2	m	km	m^3
15,000	4×10^{-5}	59,683	0.30

Table 6.2. *Characteristic of the hollow fibers contained in the filtration module*

The specific surface of the filtration module, which evaluates the filtration surface per unit of volume, is greater as the elementary size is reduced:

$$\frac{S}{Vol_{fibres}} = \frac{2}{R_e}$$

The choice of small fibers allows the filtration surface to be considerably increased.

8) The equilibrium of forces applied on the surfaces enclosing the fiber volume of arc element $2d\theta$ and length L is written as:

$$P_{int}(LR_i 2d\theta)\vec{e}_r - P_{riv}(LR_e 2d\theta)\vec{e}_r + T\{\cos(d\theta)\,\vec{e}_\theta - \sin(d\theta)\,\vec{e}_r\} \\ + T\{-\cos(d\theta)\,\vec{e}_\theta - \sin(d\theta)\,\vec{e}_r\} = \vec{0}$$

The modulus T of the tension force is calculated by considering the projection on $\vec{e}_r$. For an angle $d\theta \to 0$:

$$P_{int}(LR_i 2d\theta) - P_{riv}(LR_e 2d\theta) - 2Td\theta = 0$$

Tension is related to stretching stress σ by $T = \sigma Le$. Hence:

$$\sigma = P_{int}\frac{R_i}{e} - P_{riv}\frac{R_e}{e}$$

The results are given in Table 6.3, with $P_{int} = P_{riv} + P_{osm,int}/2$.

$P_{osm,int}$	P_{int}	P_{riv}	σ
MPa	MPa	MPa	MPa
2.26	1.23	0.1	1.03

Table 6.3. *Calculation of the stretching stress of hollow fibers in the filtration module*

Hollow fibers used in reverse osmosis are often produced in cellulose acetate or polyamide. The maximum stretching stresses (beyond which it breaks) are:

– for cellulose acetate: $\sigma_{max} \approx 10 - 110$ MPa;

– for polyamide: $\sigma_{max} \approx 80$ MPa.

It is verified that the stretching stresses are much lower than the stress causing rupture. This is obtained due to the relationships $R_i/e = 1$ and $R_e/e = 2$. It would be enough to increase the radius by a factor of 10, hence $R_e = 400$ μm, to cause the rupture.

9) The first conclusion, in terms of energy, is that the power of 0.68 kW (question 6) does not make it possible to satisfy the needs of a single house (it would rather be necessary to have an average power of 1 kW). For comparison, the same module can supply 700 houses with drinking water (see section 6.3). The handicap of the technology is therefore first and foremost economic. The investment and operating costs for producing electricity with a hollow fiber module are disproportionate to the use that can be made of it to produce drinking water.

The production of 1 m^3 of drinking water has an energy cost of 2 kWh. Its cost of €0.4 should be compared to a selling cost per m^3 of drinking water of between €3 and €5. The energy cost of desalination is actually quite low. It is therefore not surprising that the electricity supply using the same principle is weak.

The process is natural and involves renewable energies, in the sense that the river will flow into the sea and mix with sea water anyway. Some biologists, for large-scale installations, are alarmed by biological effects due to a reduction in salinity if the fresh water and sea water mixed in the plant are discharged into the sea near the coast. For the module considered, which barely supplies a house with electricity, the osmotic flow rate is modest (1.2 L/s). A very large installation would therefore be required (which seems far from being feasible, see section 6.3) so that these ecological consequences must be carefully studied.

The level of high pressure in the reactor is relative, if we compare it to a hydroelectric plant. The pressure difference between the reactor and the river corresponds to an effective height of 115 m of water column. This is an average fall height according to the classification introduced in section 5.1. The flow rate of 4.9 L/s passing through the turbine is very low compared to those usually passing through the turbine of a hydroelectric installation.

This exercise leaves aside the sizing of the turbine and the recycling pump. Using the approach of roto-dynamic machines explained in Chapter 2, the first conclusion obtained, considering the very low flow rate, is that the specific angular speeds Ω (section 2.1.7) of these machines would necessarily be very low.

We could carry out, without any particular difficulty, the dimensioning of a Pelton turbine for the osmotic installation studied in this exercise by repeating the approach of question 3 of exercise 5.2. The only way to maintain a reasonable rotation speed of the rotor is to equip the turbine with a single injector and to choose a large form factor: $D_{rot}/d_{inj} = 30$. It remains unlikely to find a Pelton turbine on the market with the characteristics obtained because these are not common turbine conditions.

The sizing of the recycling pump following the approach of question 6 of exercise 5.2 shows for its part that a recycling pump pertaining to centrifugal technology should have a flow coefficient δ well below 0.01 (Figure 2.7). These conditions go beyond the limits of use of centrifugal pumps. The solution is to be found using volumetric pumps (not covered in this work).

It is therefore difficult to find roto-dynamic machines to ensure the heads and flow rates in the hollow fiber module (Figure 6.4) considered in this exercise. Only a much larger installation, working with the same pressure in the reactor and providing much greater flow rates, would be able to operate with the conventional turbine and pumping equipment studied in this work. The state of the art (section 6.3) shows that the current level of development of osmotic energy technology does not fall within this framework.

6.3. State of the art of osmotic energy technology

The sizing carried out in exercise 6.1 seems to conclude on the weak prospects for electricity production using osmotic energy.

Table 6.4, established with the results of the exercise, compares the production of the same filtration module to provide drinking water by reverse osmosis or electricity by osmosis.

The filtration module uses fibers of very small thickness and very small diameter. They are required to ensure sufficient flow rates and to have a large filtration surface in a small volume. The sizing brought us into contact with the high level technology of manufacturing fibers and filtration modules.

The same module with 15,000 m^2 of filtration surface can supply 700 homes with drinking water, but it can only supply one home with electricity. The investment is therefore 700 times greater to use osmosis for electricity production purposes than for drinking water production purposes.

Process	Seawater desalination	Electricity production
Process	Reverse osmosis	Osmosis
Filtration surface	15,000 m^2	15,000 m^2
ΔH through the membrane	69 bar	11.25 bar
Supply	Drinking water for 700 homes (350 m^3/day)	Electricity for 1 home
Energy/power	2 kWh per m^3 of drinking water	0.68 kW

Table 6.4. *Comparison of the uses of a filtration module to provide drinking water by reverse osmosis or electricity by osmosis*

It is not very easy to find precise figures on the investment cost of reverse osmosis filtration modules. A BRGM report[3], quite old, gives the range of €1,000–€2,000 per m^3/day. For the exercise module providing 350 m^3/day, the investment cost would therefore be between €350,000 and €700,000. To provide water to 700 houses, the investment seems feasible. It is not intended to provide electricity to a single house.

Statkraft, a Norwegian company important in the energy market, well recognized for its work in the field of renewable energies, has built a prototype of energy production based on osmosis. S.-E. Skilhagen, osmotic project manager at Statkraft, declared in 2009[4]: "Our current prototype has 2,000 m^2 of spiral membranes from which we should recover 3 watts per square meter (W/m^2). To be profitable, it would be necessary to reach 5W/m^2". The same article from the newspaper *Les Échos* announced the inauguration at the end of November 2009 of the world's first prototype of an osmotic plant, in Tofte near Oslo, with power between 2 and 4 kW. The figure of 3 W/m^2 is at least 66 times more efficient than that of the hollow fibers considered in exercise 6.1. The estimated power of 0.68 kW with a filtration surface of 15,000 m^2 comes to 0.045 W/m^2, neglecting any head loss other than that linked to filtration.

We therefore measure the difference between our small exercise model and a real prototype and the progress made in the performance of the filtration fibers. Although

3 See BRGM (2003).

4 *Les Échos,* October 23, 2009.

a factor of 66 is not small, the comparison nevertheless accredits the relevance of the model as a tool for understanding the process.

In a press release published on December 20, 2013 on its website, Statkraft declared:

> Statkraft is ending its work to develop osmotic technology to produce electricity. In the current market outlook, the company recognizes that osmotic technology will not be sufficiently developed in the near future to become competitive.
>
> Our challenge was to make this technology efficient enough to produce electricity at competitive costs compared to other electricity production sectors. In market conditions, we see that this cannot be achieved in the near future.

7

Ocean Thermal Energy Conversion

7.1. The OTEC process

The ocean thermal energy conversion (OTEC) process consists of producing mechanical energy by implementing a Rankine cycle, which recovers thermal energy from a hot source of sea water and restores thermal energy to a cold seawater source. The mechanical energy supplied to the turbine activates an alternator to produce electrical energy.

The process requires having two sources of sea water nearby, one hot and the other cold. It can be implemented in tropical and equatorial zones where surface waters reach a temperature of around $T_{sea,h} \cong 30°C$, while deep waters are at a temperature of around $T_{sea,c} \cong 5°C$. Like cold water is pumped from the depths, it is important that the great depths are accessible within a short distance.

The potential sites for setting up an OTEC factory are therefore not very numerous on Earth. The island of Hawaii and Reunion Island are among them, since the mountainous terrain of these islands is associated with deep seas a short distance from the shore. For isolated areas with limited energy resources, the OTEC process can be an alternative for the supply of electricity (see section 7.3).

The temperature difference between warm sea water and cold sea water is small. As a result, the efficiency of the cycle, defined as the ratio of the mechanical power $\dot{W}_{cycle}$ supplied by the cycle to the thermal power Φ_{evap} transferred by the hot source, is low. The efficiency of the Rankine cycle is in fact lower than the efficiency of the ideal Carnot cycle:

$$r_{Rankine} = \frac{\dot{W}_{cycle}}{\Phi_{evap}} < r_{Carnot} = 1 - \frac{T_{sea,c}}{T_{sea,h}}$$

The temperatures $T_{sea,c}$ and $T_{sea,h}$ in the formula above are the absolute temperatures evaluated in Kelvin.

Figure 7.1 shows the block diagram of an installation where a fluid (ammonia – black fluid loop in the figure) undergoes a Rankine cycle. Four pieces of equipment govern the thermodynamic transformations of the cycle:

– The condenser: the ammonia is condensed there by transferring the thermal power Φ_{cond} to the cold sea water. Condensation occurs by maintaining the pressure of the ammonia in the condenser at the value of the vapor pressure $P = P_v(T_{NH3,cond})$ for the $T_{NH3,cond}$ chosen temperature. The temperature of the ammonia $T_{NH3,cond}$ in the condenser must be higher than the temperature $T_{sea,c}$ of the cold sea water. The condenser must be sized so that all of the ammonia is in liquid form leaving the condenser.

– The liquid pump: it increases the pressure of the liquid ammonia from the value $P = P_v(T_{NH3,cond})$ to the value $P = P_v(T_{NH3,evap})$. The ammonia temperature remains $T_{NH3,cond}$.

– The evaporator: the ammonia is evaporated there using the thermal power Φ_{evap} provided by the hot sea water. The ammonia evaporation temperature $T_{NH3,evap}$ is set by the vapor pressure $P = P_v(T_{NH3,evap})$ maintained in the evaporator. The temperature of the ammonia in the evaporator should be lower than the temperature $T_{sea,h}$ of the warm sea water. The evaporator must be sized so that all of the ammonia is in gaseous form leaving the evaporator.

– The turbine: the expansion of the gas from pressure $P_v(T_{NH3,evap})$ to pressure $P_v(T_{NH3,cond})$ provides mechanical power to the turbine which activates an alternator to convert into electrical power.

The first two operating parameters of the Rankine cycle are the evaporation temperature $T_{NH3,evap}$ (necessarily lower than the temperature of warm sea $T_{sea,h}$ water) and the condensation temperature $T_{NH3,cond}$ (necessarily higher than the temperature of cold sea water $T_{sea,c}$). The choice of these two temperatures sets the pressures $P_v(T_{NH3,evap})$ and $P_v(T_{NH3,cond})$ which must be maintained in the evaporator and the condenser, respectively. The variations of the vapor pressure of ammonia $P_v(T)$ with temperature are thermodynamic data.

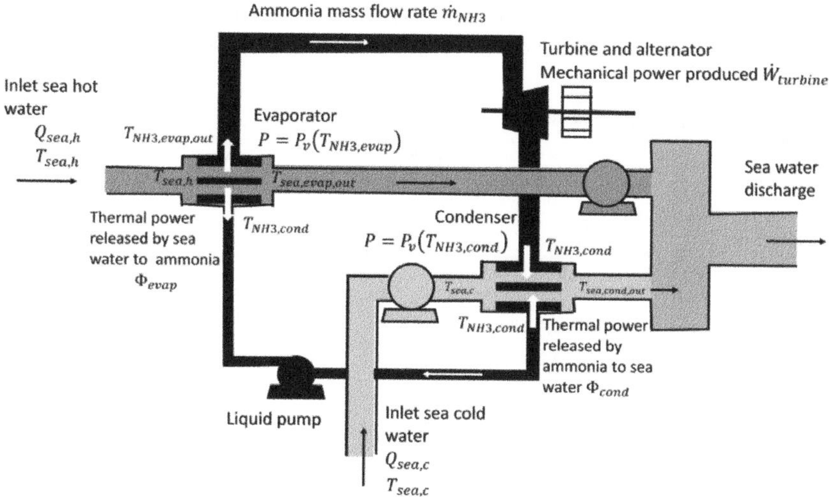

Figure 7.1. *Diagram of an OTEC installation where a fluid (black loop) undergoes a Rankine cycle*

A constraint to be respected for proper operation of the Rankine cycle is that the expansion of the gas in the turbine is isentropic, without the risk of partial condensation of the ammonia occurring. This requires that the overheating temperature $T_{NH3,evap,out}$ at the evaporator outlet (equal to the turbine inlet temperature) is linked to the turbine outlet temperature by the Laplace relation:

$$(T_{NH3,evap,out})^{\gamma}\left(P_v(T_{NH3,evap})\right)^{1-\gamma} = (T_{NH3,cond})^{\gamma}\left(P_v(T_{NH3,cond})\right)^{1-\gamma}$$

Consequently, the ammonia gas must be overheated in the evaporator to a temperature $T_{NH3,evap,out}$ above the evaporation temperature $T_{NH3,evap}$.

The turbine enthalpy balance determines the ammonia mass flow rate $\dot{m}_{NH3}$ necessary to provide the desired mechanical power $\dot{W}_{cycle}$. The thermal powers exchanged in the evaporator and in the condenser, Φ_{evap} and Φ_{cond}, are then deduced from the enthalpy balances of the evaporator and the condenser.

Figure 7.2 represents the Rankine cycle in a diagram (T, P). The variation $P_v(T)$ in the vapor pressure of ammonia is superimposed (dashed line) on this diagram on the thermodynamic cycle. The expansion curve in the turbine (DA) is below the curve $P_v(T)$, ensuring that the ammonia remains in gaseous form in the turbine. In

Figure 7.2, the different stages of the energy conversion processes at work in the OTEC process using a closed Rankine cycle are indicated in bold.

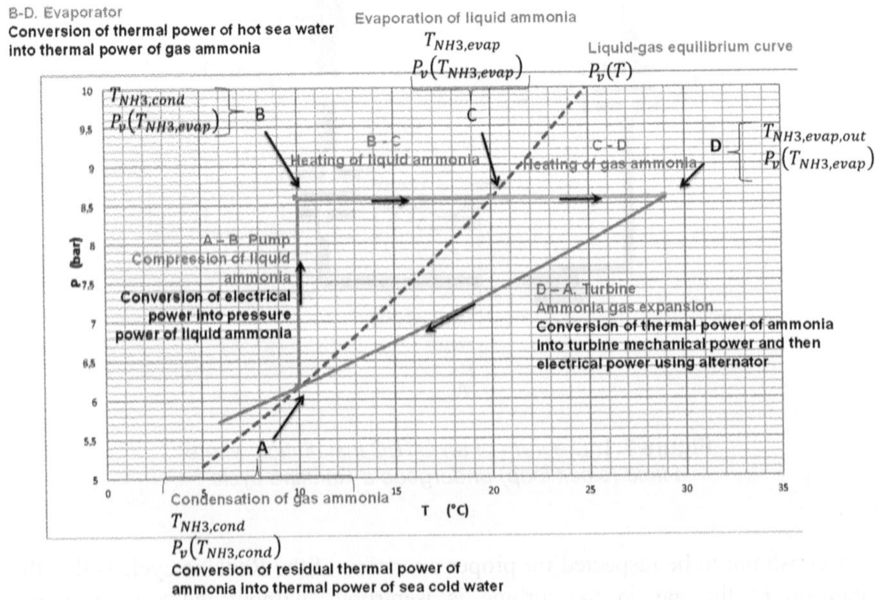

Figure 7.2. *Representation of the Rankine cycle in a diagram (T, P). For a color version of this figure, see www.iste.co.uk/mory/mechanical.zip*

The exercise proposed in section 7.2 studies the sizing of a 1 MW OTEC installation using a closed Rankine cycle. This sizing is a compromise between the choice of the operating conditions of the Rankine cycle and the choice of the operating conditions of the two exchangers (evaporator and condenser).

Choosing the evaporation temperature $T_{NH3,evap}$ as close as possible to the hot seawater temperature $T_{sea,h}$ and simultaneously choosing the condensation temperature $T_{NH3,cond}$ as close as possible to the cold seawater temperature $T_{sea,c}$ helps minimize the mass flow rate of ammonia $\dot{m}_{NH3}$ required to provide mechanical power $\dot{W}_{cycle}$.

This choice reduces the temperature differences between the water and the ammonia in the two exchangers. To transfer the required thermal powers Φ_{evap} and Φ_{cond}, it will then be necessary to increase the exchange surfaces of the exchangers and, consequently, their dimensions.

To solve the exercise in section 7.2, we can use the guide on the sizing of exchangers provided in section 7.4.

Section 7.3 presents an overview of the OTEC sector.

7.2. Exercise: sizing an OTEC installation using a closed Rankine cycle

We size in this exercise an energy production installation of power $\dot{W}_{cycle}$ = 1 MW using thermal energy from the sea. The process implements a Rankine cycle where ammonia circulates in a closed loop. The hot source for evaporating the ammonia is surface seawater at the temperature $T_{sea,h} = 32°C$ and the cold source for subsequently condensing it is seawater pumped deep from the sea, the temperature of which is $T_{sea,c} = 5°C$. The block diagram of the process has been presented in Figure 7.1.

The exercise has three parts. In the first, the Rankine cycle is dimensioned and the mass flow rate of ammonia $\dot{m}_{NH3}$ necessary for the cycle to provide mechanical power $\dot{W}_{cycle} = 1MW$ is determined. The second part is devoted to the sizing of the heat exchangers, evaporator and condenser. The third part studies the pumping of cold water from the deep layers of the sea and summarizes the conclusions of the exercise.

7.2.1. *Sizing of the closed Rankine cycle*

The Rankine cycle is sized by choosing the evaporation temperature $T_{NH3,evap} = 20°C$ and the condensation temperature $T_{NH3,cond} = 10°C$ of the ammonia. It is broken down into four transformations (Figure 7.2):

– isothermal compression (A-B) of ammonia by a pump which increases the pressure of liquid ammonia from vapor pressure $P_v(T_{NH3,cond})$ to vapor pressure $P_v(T_{NH3,evap})$;

– heating and evaporation (B-D) of ammonia. In the evaporator, the pressure is maintained at the value $P_v(T_{NH3,evap})$. Liquid ammonia is first heated by warm sea water to change from $T_{NH3,cond}$ to $T_{NH3,evap}$. It then evaporates at temperature $T_{NH3,evap}$. The gas then undergoes overheating to reach the overheating temperature $T_{NH3,evap,out}$, which will be calculated during sizing;

– isentropic expansion (D-A) of ammonia gas in the gas turbine, which lowers the temperature of the ammonia gas from $T_{NH3,evap,out}$ to $T_{NH3,cond}$. The mechanical energy recovered by the turbine is used to drive the alternator and produce electricity;

– condensation of ammonia (A-A) in the condenser. The temperature in the condenser is maintained at the temperature $T_{NH3,cond}$ and the pressure at the value $P_v(T_{NH3,cond})$.

1) The thermodynamic data on ammonia necessary for the design of the Rankine cycle are given in Table 7.1.

Temperature	$P_v(T)$	Gas energy $h_{u,g}$	Gas enthalpy $h_{p,g}$	Liquid enthalpy $h_{p,l}$
°C	Bar	kJ/kg	kJ/kg	kJ/kg
10	6.15	382.53	508.89	– 716.14
20	8.58	389.06	517.01	– 668.82

Table 7.1. *Thermodynamic data of ammonia*[1]

Suppose that ammonia behaves like an ideal gas and that the variations in specific heats $c_{v,NH3,g}$, $c_{p,NH3,l}$ and $c_{p,NH3,g}$ are negligible in the temperature intervals considered; determine:

– specific heats $c_{v,NH3,g}$, $c_{p,NH3,g}$ and $c_{p,NH3,l}$;

– the Laplace coefficient $\gamma = c_{p,NH3,g}/c_{v,NH3,g}$;

– the latent heats of evaporation and condensation at Rankine cycle temperatures.

Neglecting the variations in density of liquid ammonia with temperature, we will take its value equal to $\rho_{NH3,l} = 625 \text{ kg/m}^3$.

2) We dimension the Rankine cycle in ideal conditions where the efficiencies of the pump and the turbine are equal to 1. Calculate the overheating temperature $T_{NH3,evap,out}$ which must be reached in the evaporator, so as to ensure isentropic expansion in the turbine and avoid ammonia condensation. Represent the Rankine cycle on a diagram (*P*,*T*) and indicate on it the values of the temperature and pressure for the remarkable points of the cycle.

3) Calculate the mechanical power required for the pump to compress a unit mass flow rate of liquid ammonia $\dot{m}_{NH3} = 1 \text{ kg/s}$ during transformation (A-B) of the cycle.

1 See Haar and Gallagher (1978).

4) Calculate the thermal power that the hot sea water must provide in the evaporator to heat the liquid ammonia, evaporate it and then overheat it in the gas phase, for a unit mass flow rate $\dot{m}_{NH3} = 1\ \mathrm{kg/s}$ of ammonia (transformation B-D of the cycle).

5) Calculate the mechanical power recovered by the turbine during the isentropic expansion of ammonia gas (transformation D-A) for a unit mass flow rate $\dot{m}_{NH3} = 1\ \mathrm{kg/s}$.

6) Calculate the thermal power supplied to cold water by ammonia during its condensation (transformation A-A) for a unit mass flow rate of ammonia $\dot{m}_{NH3} = 1\ \mathrm{kg/s}$.

7) Calculate the mechanical power supplied by the cycle for a unit mass flow rate $\dot{m}_{NH3} = 1\ \mathrm{kg/s}$ of ammonia. Calculate the efficiency of the cycle and compare it to that of a Carnot cycle using the same hot and cold sources.

8) Determine the mass flow rate of ammonia $\dot{m}_{NH3}$ for which the Rankine cycle provides power $\dot{W}_{cycle} = 1$ MW. For this mass flow rate of ammonia, summarize in a table the mechanical powers consumed by the pump and recovered by the turbine as well as the thermal powers exchanged in the evaporator and in the condenser.

7.2.1.1. *Solution*

1) Assuming ammonia is an ideal gas, its energy and enthalpy depend only on temperature. We can therefore use the values of internal energy and enthalpy given in Table 7.1 to calculate the specific heats in the gas phase:

$$c_{p,NH3,g} = \frac{h_{p,g}\left(T_{NH3,evap}\right) - h_{p,g}\left(T_{NH3,cond}\right)}{T_{NH3,evap} - T_{NH3,cond}}$$

$$c_{v,NH3,g} = \frac{h_{u,g}\left(T_{NH3,evap}\right) - h_{u,g}\left(T_{NH3,cond}\right)}{T_{NH3,evap} - T_{NH3,cond}}$$

The Laplace coefficient $\gamma = c_{p,NH3,g}/c_{v,NH3,g}$ is deduced from these values.

Neglecting variations in volume of liquid ammonia, the variation in enthalpy produced by a variation in temperature ΔT and pressure ΔP is written as:

$$h_{p,l}(T + \Delta T, P + \Delta P) - h_{p,l}(T, P) = c_{p,l}\Delta T + \frac{\Delta P}{\rho_l}$$

We can therefore estimate the specific heat $c_{p,NH3,l}$ from the enthalpy data of liquid ammonia for both temperature and pressure values.

Finally, at a given temperature, the latent heat of phase change is:

$$L_v(T) = h_{p,g}(T, P_v(T)) - h_{p,l}(T, P_v(T))$$

The numerical values are given in Table 7.2[2].

<table>
<tr><td>T_{NH3}</td><td>P_v</td><td>$h_{p,l}$</td><td>$c_{p,l}$</td><td>$h_{p,g}$</td><td>$c_{p,g}$</td></tr>
<tr><td>°C</td><td>Bar</td><td>kJ/kg</td><td>kJ/kg/K</td><td>kJ/kg</td><td>kJ/kg/K</td></tr>
<tr><td>10</td><td>6.15</td><td>– 716.14</td><td rowspan="2">4.69</td><td>508.89</td><td rowspan="2">0.812</td></tr>
<tr><td>20</td><td>8.58</td><td>– 668.82</td><td>517.01</td></tr>
<tr><td>T_{NH3}</td><td>P_v</td><td>$h_{u,g}$</td><td>$c_{v,g}$</td><td>γ</td><td>L_v</td></tr>
<tr><td>°C</td><td>Bar</td><td>kJ/kg</td><td>kJ/kg/K</td><td>–</td><td>kJ/kg</td></tr>
<tr><td>10</td><td>6.15</td><td>382.53</td><td rowspan="2">0.653</td><td rowspan="2">1.24</td><td>1,225</td></tr>
<tr><td>20</td><td>8.58</td><td>389.06</td><td>1,186</td></tr>
</table>

Table 7.2. *Numerical values of specific heats, Laplace coefficient and latent heats at the condensation and vaporization temperatures of ammonia*

2) When the expansion of ammonia gas is isentropic in the turbine, the temperatures and pressures at the inlet and outlet of the turbine are linked by Laplace's law, from which we deduce the overheating temperature $T_{NH3,evap,out}$ which must be reached at the evaporator outlet:

$$T_{NH3,evap,out} = T_{NH3,cond}\left(\frac{P_v(T_{NH3,evap})}{P_v(T_{NH3,cond})}\right)^{(\gamma-1)/\gamma}$$

Temperatures must be expressed in Kelvin in this formula. With the thermodynamic data calculated in the previous question, the overheating temperature is worth $T_{NH3,evap,out} = 29.0$°C.

The numerical values obtained confirm that the Rankine cycle diagram (P, T) presented in Figure 7.2 corresponds to the data from this exercise. On this graph, the superimposed variation of vapor pressure with temperature indicates that ammonia remains in gaseous form during gas expansion. If the overheat temperature

2 In this and subsequent tables in this exercise, cells are unshaded when they contain data and shaded when they contain results.

were less, the isentropic curve would intersect the curve $P_V(T)$ before reaching the condensation temperature, allowing gas condensation to occur inside the turbine. In the evaporator, the temperature increases from 10°C to 29°C. The heat is provided by the hot sea water which enters at a temperature of 32°C. In the condenser the temperature is 10°C, higher than the cold water inlet temperature (5°C) at which the ammonia gives up heat.

3) The mechanical power required by the pump to increase the pressure of liquid ammonia from pressure $P_v(T_{NH3,cond})$ to pressure $P_v(T_{NH3,evap})$, while temperature remains at the value $T_{NH3,cond}$, is determined, neglecting the variation in density of liquid ammonia, by calculating the enthalpy change:

$$\dot{W}_{pump} = \dot{m}_{NH3}\left\{\frac{P_v(T_{NH3,evap}) - P_v(T_{NH3,cond})}{\rho_{NH3,l}}\right\}$$

The numerical results from questions 3 to 7 will be summarized in Table 7.3 in question 8.

4) The thermal power provided by the hot sea water in the evaporator to heat the liquid ammonia up to the evaporation temperature, ensure evaporation and then heat the gas ammonia up to the overheating temperature is:

$$\Phi_{evap} = \dot{m}_{NH3}\{c_{p,NH3,l}(T_{NH3,evap} - T_{NH3,cond}) + L_{v,NH3}(T_{NH3,evap}) + c_{p,NH3,g}(T_{NH3,evap,out} - T_{NH3,evap})\}$$

5) Assuming the operation of the turbine is adiabatic, the mechanical power recovered by the turbine is deduced from the variation in enthalpy between the limit points of the transformation D-A:

$$\dot{W}_{turbine} = \dot{m}_{NH3}\{h_{p,g}(T_{NH3,evap,out}) - h_{p,g}(T_{NH3,cond})\} = \dot{m}_{NH3}c_{p,NH3,g}(T_{NH3,evap,out} - T_{NH3,cond})$$

6) In the condenser, the ammonia in gaseous form is completely condensed at temperature $T_{NH3,cond}$. The thermal power transferred to cold sea water is:

$$\Phi_{cond} = \dot{m}_{NH3}L_{v,NH3}(T_{NH3,cond})$$

7) The mechanical power supplied by the cycle is:

$$\dot{W}_{cycle} = \dot{W}_{turbine} - \dot{W}_{pump}$$

and the efficiency of the cycle is:

$$r_{Rankine} = \frac{\dot{W}_{cycle}}{\Phi_{evap}}$$

A Carnot cycle operating with the same hot and cold sources has the yield (temperatures in K):

$$r_{Carnot} = 1 - \frac{T_{sea,c}}{T_{sea,h}}$$

8) We summarize in Table 7.3 the mechanical and thermal powers exchanged during the Rankine cycle in two cases: where the mass flow rate is $\dot{m}_{NH3} = 1$ kg/s and that for which the power supplied by the cycle is $\dot{W}_{cycle} = 1$ MW.

$\dot{m}_{NH3}$	$\dot{W}_{pump}$	Φ_{evap}	$\dot{W}_{turbine}$	–	$\dot{W}_{cycle}$	$r_{Rankine}$	r_{Carnot}
kg/s	kW	kW	kW	kW	kW	–	–
1	0.39	1,240	12.54	1,225	12.15	0.010	0.089
82.3	31.9	102,076	1,031.9	100,837	1,000		

Table 7.3. *Thermal and mechanical powers supplied and consumed in the four transformations of the Rankine cycle, for an ammonia flow rate of 1 kg/s and for a mechanical power supplied by the cycle of 1,000 MW*

7.2.2. *Sizing of heat exchangers*

The heat exchangers must transfer the thermal power Φ_{evap} from the hot seawater ($T_{sea,h} = 32°C$) to the ammonia in the evaporator and the power Φ_{cond} from the ammonia to the cold seawater ($T_{sea,c} = 5°C$) in the condenser. The numerical values of the required thermal powers were given in question 8. The sizing of the exchangers consists of determining their geometry (diameter, length and number of tubes, transfer surface) and choosing the cold water and hot water flow rates.

The physical properties of water and ammonia complementary to those already given in section 7.2.1 appear in Table 7.4.

We can use the summary in section 7.4 for heat exchanger sizing to carry out the dimensions of the evaporator and the condenser.

Setting	Symbol	Unit	Liquid water H_2O	Liquid ammonia NH_3	Gas ammonia NH_3
Volumic mass	ρ	kg/m^3	1,000	625	–
Massic heat	c_p	J/kg/K	4,180	–	–
Molar mass	M	g/mol	18	17	17
Thermal conductivity	Λ	W/m/K	0.60	0.49	0.0265
Dynamic viscosity	μ	Pa.s	10^{-3}	1.3×10^{-4}	10^{-5}

Table 7.4. *Thermodynamic data of ammonia and water*

7.2.2.1. *Evaporator sizing*

A schematic of the evaporator is shown in Figure 7.3. This is a "tube and shell" type heat exchanger. The hot fluid, warm sea water entering at temperature $T_{H2O,in} = T_{sea,h}$, flows inside the tubes of diameter $D = 4$ cm spaced regularly at a distance $d = 5$ cm. The number n^2 of tubes contained in the shell is one of the sizing parameters. The ammonia flows into the shell, between the tubes. The exchanger is countercurrent. The ammonia inlet is opposite the water outlet and vice versa. The ammonia inlet and outlet temperatures are set by the Rankine cycle. The water outlet temperature $T_{H2O,out}$ depends on the water flow $Q_{H2O,evap}$. The choice of one of these two quantities determines the value of the other. Countercurrent operation makes it possible to have an identical temperature difference between the two fluids at both ends of the exchanger and thus maintain more homogeneous heat exchanges across the surface S through which the heat is transferred from the hot seawater to the ammonia.

Since the water transfers heat to the ammonia, its temperature decreases continuously from inlet to outlet. On the other hand, as the ammonia evaporates, its temperature remains equal to the evaporation temperature $T_{NH3,evap}$ in a large portion of the exchanger, as shown schematically by the temperature profiles in Figure 7.3.

In steady state, the thermal power exchanged in the evaporator is expressed in three forms: as the power recovered by the ammonia (question 4), as the power transferred by sea water, or as the power exchanged between both fluids. The last two forms are written as:

$$\Phi_{evap} = Q_{H2O,evap}\rho_{H2O,l}c_{p,H2O,l}\left(T_{H2O,in} - T_{H2O,out}\right) = K.S.\Delta T$$

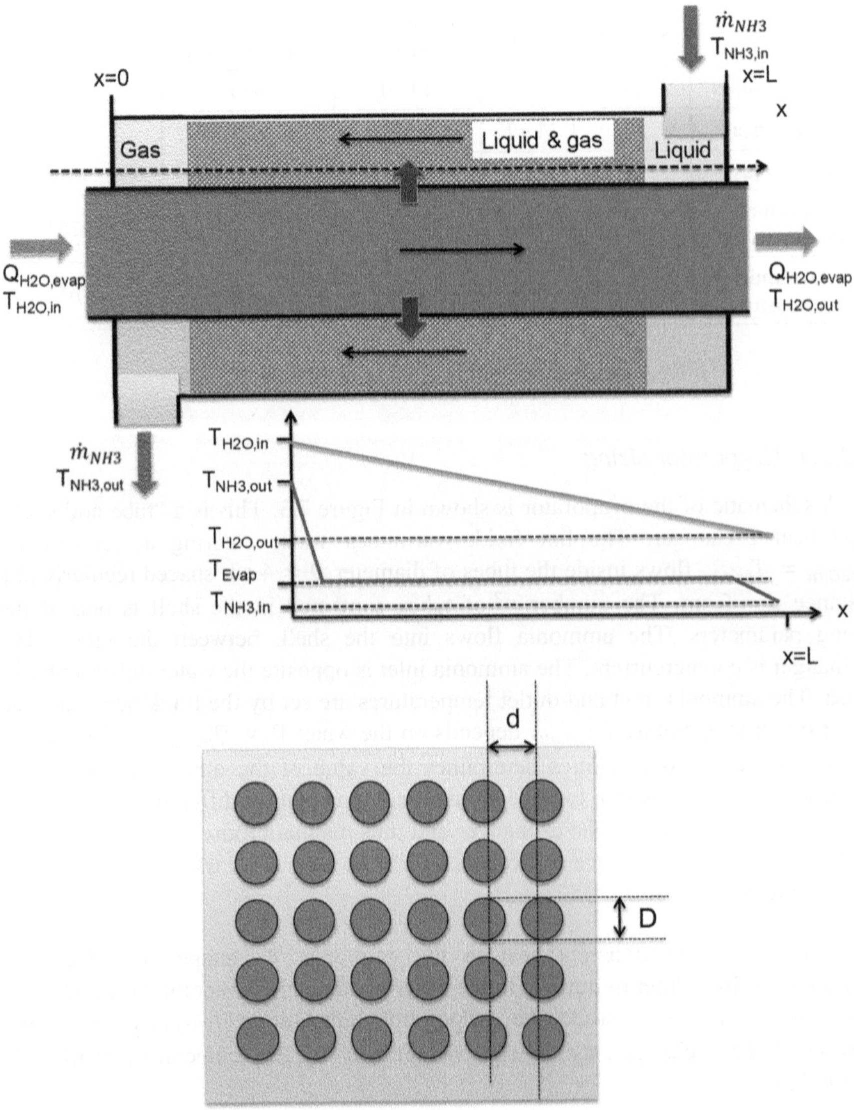

Figure 7.3. *Countercurrent evaporator diagram and temperature profiles. For a color version of this figure, see www.iste.co.uk/mory/mechanical.zip*

In the last expression, K is the overall transfer coefficient between the two fluids, S the surface area through which the heat is transferred and ΔT is the average temperature difference between the water and the ammonia.

9) Summarize in a table the values of temperatures $T_{H2O,in}$, $T_{NH3,in}$ and $T_{NH3,out}$. Determine the temperature $T_{H2O,out}$ for which the temperature difference between the two fluids is the same at both ends of the exchanger. Calculate the water flow rate $Q_{H2O,evap}$ required to obtain this outlet water temperature. In this case, the average temperature difference ΔT between the two fluids in the exchanger is equal to what it is at the ends.

10) The overall heat transfer coefficient K of the evaporator combines the partial heat transfer coefficients in water and in ammonia:

$$K = \frac{h_{H2O} h_{NH3}}{h_{H2O} + h_{NH3}}$$

Propose a value of n for which n^2 is the number of tubes contained in the shell. Adjust the value of n in such a way that the flow speed of water in the exchanger is in the interval $0.3\ \mathrm{m/s} < U_{H2O} < 0.5\ \mathrm{m/s}$.

Since the water remains in liquid form and the transfer regime is that of forced convection, calculate the partial heat transfer coefficient h_{H2O} in water using the Colburn correlation in turbulent regime:

$$h_{H2O} = \frac{\Lambda_{H2O,l}}{D_H} 0.023 Pr_{H2O}^{1/3} Re_{H2O}^{0.8}$$

with $Re_{H2O} = \frac{\rho_{H2O,l} U_{H2O} D_H}{\mu_{H2O,l}} > 10{,}000$ and $Pr_{H2O} = \frac{\mu_{H2O,l} c_{p,H2O,l}}{\Lambda_{H2O,l}}$.

The hydraulic diameter D_H is proportional to the ratio of the flow cross-section to the perimeter of the line surrounding this surface. For the present configuration with circular tubes, $D_H = D$. We will verify that the Reynolds number is in the range of the turbulent regime.

11) Due to the phase change of ammonia, the precise determination of the partial transfer coefficient of ammonia h_{NH3} is difficult. Considering that the transfer coefficients in an evaporating fluid are generally much higher than those in a single-phase fluid, we will take $h_{NH3} = 10{,}000\ \mathrm{W/m^2/K}$. Calculate the overall heat transfer K coefficient and check that its value varies little when the value of h_{NH3} is varied in the interval usually considered (Table 7.15, section 7.4) in an evaporating fluid. For the value n chosen in question 10, calculate the transfer surface S and the length of tubes L. Finally, calculate the volume of the shell.

12) Calculate the linear pressure loss of the water flow in the evaporator and the power required to ensure this flow[3].

Calculate the speed of ammonia in the evaporator when it is completely in the liquid phase or in the gas phase. Use these values to argue that the power required to maintain the flow of ammonia in the exchanger is low.

13) If using a spreadsheet to carry out the numerical applications of this exercise, we recommend varying the value of n to see how it influences the transfer coefficients and head losses and to understand the suggested speed interval in question 10 to carry out the sizing.

7.2.2.2. *Condenser sizing*

The condenser (Figure 7.4) is a shell and tube type countercurrent exchanger. The diameter of the tubes and their spacing are identical to the values taken for the evaporator ($D = 4$ cm, $d = 5$ cm.). As with the evaporator, sizing consists of choosing an integer n setting the number of tubes n^2 in the exchanger.

Since ammonia is the hot fluid, it is now flowing through the tubes while the water flows through the shell.

14) Summarize in a table the values of temperatures $T_{H2O,in}$, $T_{NH3,in}$ and $T_{NH3,out}$. Note that the temperature difference between ammonia and water cannot be identical at both ends of the exchanger. Calculate water flow rate $Q_{H2O,cond}$ ensuring that the outlet water temperature $T_{H2O,out}$ is 2°C below the ammonia temperature $T_{NH3,in}$. According to the theory of exchangers (see section 7.4), the average temperature difference between the two fluids in the exchanger is calculated using the logarithmic average:

$$\Delta T_{mLn} = \frac{\left(T_{NH3,in} - T_{H2O,out}\right) - \left(T_{NH3,out} - T_{H2O,in}\right)}{Ln\left(\frac{T_{NH3,in} - T_{H2O,out}}{T_{NH3,out} - T_{H2O,in}}\right)}$$

15) Express as a function of the integer n and the lengths d and D the flow cross section $S_{Q,H2O}$ in the shell. Derive the expression for the hydraulic diameter $D_H = 4S_{Q,H2O}/P_{Q,H2O}$, where $P_{Q,H2O}$ is the perimeter of the lines limiting the surface $S_{Q,H2O}$. In Figure 7.4, $S_{Q,H2O}$ is the blue surface surrounding the tubes whose surface is orange, while $P_{Q,H2O}$ is the perimeter of the n^2 dark gray tubes to which is added the square perimeter of the shell.

---

3 The linear head loss coefficient and the thermal conductivity are often designated by the same symbol λ. Since these two parameters are involved in this exercise, we have designated the thermal conductivity by the symbol Λ to avoid confusing it with the linear head loss coefficient.

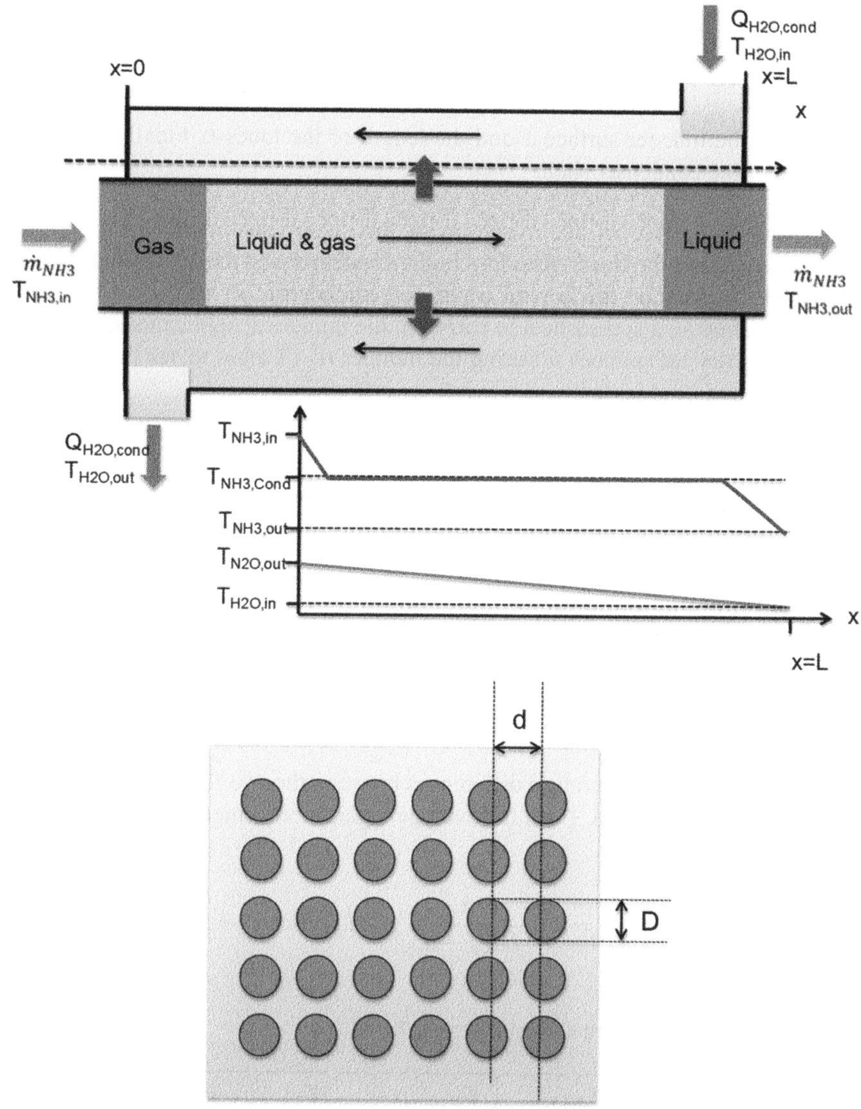

Figure 7.4. *Countercurrent condenser diagram and temperature profiles. For a color version of this figure, see www.iste.co.uk/mory/mechanical.zip*

Propose a value of n for which the flow speed of the water in the exchanger is in the interval $0.3 \text{ m/s} < U_{H2O} < 0.5 \text{ m/s}$. For this value, calculate the hydraulic diameter D_H and the Reynolds number Re_{H2O}. Using the Colburn formula, calculate the partial heat transfer coefficient h_{H2O}, then the overall transfer coefficient K. We will assume that the partial transfer coefficient in ammonia is h_{NH3} = 70,000 W/m²/K (Table 7.15, section 7.4).

Calculate the transfer surface S and the length of the tubes L. Finally, calculate the volume of the condenser shell.

16) Using the value of the linear head loss coefficient determined for the evaporator, calculate the linear head loss in the condenser and the power that a pump must provide to maintain the flow of water in the condenser. If we have established a calculation sheet in a spreadsheet to carry out the numerical applications, it will be interesting to vary the number n setting the number n^2 of tubes to see the effect on heat transfers and head losses.

7.2.2.3. *Solution*

9) The water outlet temperature is deduced from the choice made in the statement:

$$\Delta T_1 = T_{H2O,in} - T_{NH3,out} = \Delta T_2 = T_{H2O,out} - T_{NH3,in}$$

The water flow rate in the evaporator is then deduced from the thermal power:

$$\Phi_{evap} = Q_{H2O,evap}\rho_{H2O,l}c_{p,H2O,l}\left(T_{H2O,in} - T_{H2O,out}\right)$$

For a small gap in temperature differences between the two fluids at both ends of the exchanger, $\Delta T_1 = \Delta T_2(1+x)$, the average logarithmic temperature difference is given as:

$$\Delta T_{mLn} = \frac{\Delta T_1 - \Delta T_2}{Ln\left(\frac{\Delta T_1}{\Delta T_2}\right)} = \Delta T_2 \frac{x}{Ln(1+x)} \rightarrow \Delta T_2 \text{ when } x \rightarrow 0$$

The numerical results are shown in Table 7.5.

$T_{H2O,in}$	$T_{NH3,in}$	$T_{NH3,out}$	$T_{H2O,out}$	ΔT_1	Φ_{evap}	$Q_{H2O,evap}$
°C	°C	°C	°C	°C	kW	m³/s
32	10	29.0	13.0	3.0	102,076	1.28

Table 7.5. *Evaporator operating parameters and water flow rate determination*

10) Water, which is the hot fluid, flows inside the n^2 circular cylindrical tubes. The flow surface is $S_{Q,H2O} = n^2(\pi D^2/4)$ and the flow speed is $U_{H2O} = Q_{H2O,evap}/S_{Q,H2O}$.

Choosing a value $n = 50$, that is, $n^2 = 2{,}500$ tubes, allows the speed of water in the tubes to be in the interval $0.3\ \mathrm{m/s} < U_{H2O} < 0.5\ \mathrm{m/s}$. The Reynolds number is greater than 10^4 so that the partial heat transfer coefficient in water can be calculated using the Colburn formula in turbulent regime. The results are given in Table 7.6.

$Q_{H2O,evap}$	n	n^2	U_{H2O}	Re_{H2O}	$Pr_{H2O,l}$	h_{H2O}
m³/s	–	–	m/s	–	–	W/m²/K
1.28	50	2,500	0.409	16,353	6.97	1,538

Table 7.6. *Determination of the partial heat transfer coefficient in water within the evaporator*

11) Table 7.7 gives the evaluations of the overall heat transfer coefficient K for the value of the partial heat transfer coefficient in water h_{H2O} calculated in question 10 and three values of the heat transfer coefficient h_{NH3} for ammonia in boiling conditions. The transfer surface S is deduced from the expression of the thermal power:

$$\Phi_{evap} = K.S.\Delta T_{mLn}$$

We then deduce the length of the tubes:

$$L = \frac{S}{n^2 \pi D}$$

and the volume of the evaporator shell:

$$V_{evap} = L(nd)^2$$

The values given in Table 7.7 confirm that the partial heat transfer coefficient h_{NH3} in ammonia has a limited effect on the overall heat transfer coefficient compared to the variation interval.

The most important observation is the very large volume of the exchanger for the mechanical power produced by the installation $\dot{W}_{cycle} = 1$ MW. Remember that, due to the low efficiency of the cycle for the temperatures of the hot and cold seawater sources, the thermal power exchanged in the exchanger is worth 102.08 MW.

h_{H2O}	h_{NH3}	K	Φ_{evap}	ΔT_{mLn}	S	n^2	L	V_{evap}
W/m²/ K	W/m²/ K	W/m²/ K	MW	K	m²	–	m	m³
1,538	3,000	1,017	102.08	3.0	33,617	2,500	107.0	669
1,538	10,000	1,333	102.08	3.0	25,643	2,500	81.6	510
1,538	50,000	1,492	102.08	3.0	22,909	2,500	72.9	456

Table 7.7. *Calculation of the overall heat transfer coefficient in the evaporator and of the evaporator dimensions*

12) Assuming the exchanger tubes are made of slightly rusted steel, choosing a wall roughness $e = 0.2$ mm, the roughness coefficient is $k = \frac{e}{D} = 5 \times 10^{-3}$ and the linear head loss coefficient[4] is $\lambda = 0.03$ for a Reynolds number $Re_{H2O} = 16{,}300$.

The linear head loss and the power that a pump must provide to maintain the flow of water in the exchanger are given by the following formulas:

$$\Delta H_{H2O,evap} = \lambda\rho\frac{U_{H2O}^2}{2}\frac{L}{D} \quad \text{and} \quad \wp_{H2O,evap} = \Delta H_{H2O,evap} \cdot Q_{H2O,evap}$$

The flow surface in the shell is:

$$S_{Q,NH3} = n^2\left(d^2 - \frac{\pi D^2}{4}\right)$$

The flow velocity of ammonia in the liquid phase in the shell is then:

$$U_{NH3,l} = \frac{\dot{m}_{NH3}}{\rho_{NH3,l} S_{Q,NH3}}$$

For ammonia in liquid phase $\rho_{NH3,l} = 6.25$ kg/m³. In the gas phase, the density is deduced from the ideal gas law, where $M_{NH3} = 17$g/mol is the molar mass of ammonia:

$$\rho_{NH3,g} = M_{NH3}\frac{P_v(T_{evap})}{R\,T_{evap}}$$

The results are shown in Table 7.8.

4 See section 1.4.

n^2	$Q_{H2O,evap}$	U_{H2O}	λ	$\Delta H_{H2O,evap}$	$\wp_{H2O,evap}$
–	m³/s	m/s	–	Pa	kW
2,500	1.28	0.409	0.03	5,116	6.57
n^2	$\dot{m}_{NH3}$	$S_{Q,NH3}$	$U_{NH3,l}$	$\rho_{NH3,g}$	$U_{NH3,g}$
–	kg/s	m²	m/s	kg/m³	m/s
2,500	82.3	3.11	0.042	6.19	4.28

Table 7.8. *Head loss of water flow in the evaporator and flow speeds of ammonia in liquid phase and gas phase*

The power $\wp_{H2O,evap} = 6.57$ kW required to maintain the water flow through the exchanger is low compared to the power of 1 MW provided by the installation.

For ammonia, the calculation of the linear head loss is more complex since the flow is two-phase. However, the speed of the ammonia is very low in the first part of the exchanger when the ammonia is in liquid form. At the outlet, the speed is quite high but the density is low. It is therefore reasonable to think that the head losses caused by the flow of ammonia in the exchanger are also low.

13) Reducing the number of tubes by a factor of 5 (hereinafter, the calculation is made for an exchanger with 400 tubes, compared to 2,500 previously) increases the speed by the same factor. The transfer coefficient is increased, making it possible to reduce the volume of the evaporator. But the head loss and the power required to maintain the water flow increase considerably, mainly because of the increase in flow speed, secondarily because the length of the tubes also increases.

n^2	U_{H2O}	h_{H2O}	K	S	L	V_{evap}	$\Delta H_{H2O,evap}$	$\wp_{H2O,evap}$
–	m/s	W/m²/K	W/m²/K	m²	m	m³	Pa	kW
400	2.56	6,662	3,998	8,548	170.1	170.1	4.16×10^5	534.7

Table 7.9. *Effect of a reduction in the number of tubes in the evaporator on the speed of the water flow, the overall heat transfer coefficient, the dimensions of the exchanger and the head loss of the water flow*

This observation justifies choosing the number of tubes with regard to the desired speed (question 10), sufficiently large to be in turbulent regime and thus benefit from a significant heat transfer coefficient (turbulent regime), but not too much, in order to control head losses.

14) Taking the approach of question 9, the results appear in Table 7.10. We notice that the flow of cold water passing through the condenser is greater than that passing through the evaporator.

$T_{H2O,in}$	$T_{NH3,in}$	$T_{NH3,out}$	$T_{H2O,out}$	ΔT_{mLn}	Φ_{cond}	$Q_{H2O,cond}$
°C	°C	°C	°C	°C	kW	m³ /s
5	10	10	8	3.3	100,837	8.04

Table 7.10. *Condenser operating parameters and water flow determination*

15) The expression for the flow surface in the shell has already been established in question 12, but the formula now applies to water:

$$S_{Q,H2O} = n^2\left(d^2 - \frac{\pi D^2}{4}\right)$$

The length of the lines surrounding this surface is the perimeter of the n^2 tubes to which is added the perimeter of the square shell:

$$P_{Q,\text{H2O}} = 4nd + n^2\pi D$$

We deduce from this the hydraulic diameter $D_{H,H2O} = 4S_{Q,H2O}/P_{Q,H2O}$ used to calculate the partial heat transfer coefficient in water h_{H2O} and the linear head losses. The rest of the calculations then take place as in question 11. The results are given in Table 7.11.

n	n^2	$S_{Q,H2O}$	$Q_{H2O,cond}$	U_{H2O}	$P_{Q,H2O}$	$D_{H,H2O}$	Re_{H2O}
–	–	m²	m³/s	m/s	m	m	–
115	13,225	16.44	8.04	0.489	1,685	0.039	19,090
h_{H2O}	h_{NH3}	K	ΔT_{mLn}	Φ_{cond}	S	L	V_{cond}
W/m²/K	W/m²/K	W/m²/K	°C	kW	m²	m	m³
1,783	70,000	1,739	3.3	100,837	17,710	10.7	352

Table 7.11. *Calculation of the overall heat transfer coefficient and dimensions of the condenser depending on the number of tubes*

Compared with the evaporator, the number of tubes in the condenser is much larger but their length is shorter.

16) The calculations are identical to those carried out in question 12. The results, given in Table 7.12, show that the sizing controls the level of head losses in the condenser.

λ	$Q_{H20,cond}$	U_{H20}	$\Delta H_{H20,cond}$	$\wp_{H20,cond}$
–	m³/s	m/s	Pa	kW
0.03	8.04	0.489	979	7.87

Table 7.12. *Calculation of the head loss of water flow in the condenser*

7.2.3. *Sizing of deep sea water pumping and conclusions*

17) It is assumed that cold water at temperature $T_{sea,c} = 5°C$ is available at a depth of 2,000 m and that the coast is steep enough for this depth to be reached by a tube of length $L = 5{,}000$ m. If the pump used to pump cold water is placed at the shore (altitude $z = 0$) and if the water is to enter the condenser at atmospheric pressure, establish the expressions to calculate the flow velocity in the pumping line, the pressure at the pump inlet, the head and the power of the pump depending on the diameter D_{asp} of the suction pipe. Propose a value D_{asp} which protects the pumping of cold water from the risk of cavitation and for which the power of the pump is reasonable in relation to the power $\dot{W}_{cycle} = 1$ MW supplied by the unit.

18) Summarize in a table the essential characteristics of the process (powers involved in the Rankine cycle, flow rates, length of tubes and volume in the exchangers) to draw conclusions from the sizing exercise of an OTEC unit. Also discuss issues relating to the safety of the installation and its potential environmental impacts.

7.2.3.1. *Solution*

17) The flow velocity in the suction line is:

$$U_{asp} = \frac{Q_{H20,cond}}{\frac{\pi D_{asp}^2}{4}}$$

If the outlet pump pressure, at shore altitude z = 0, is atmospheric pressure and the discharge and suction pipe diameters are equal, using the generalized Bernoulli equation (section 1.6) on a current line coming from the depths and leaving the condenser (Figure 7.1), we establish that the head that the pump must provide is:

$$\Delta H_{asp} = \rho \frac{U_{asp}^2}{2}\left(1 + \lambda \frac{L}{D_{asp}}\right) + \Delta H_{H20,cond}$$

and the pressure at the pump inlet is:

$$P_{asp} = P_{atm} - \rho \frac{U_{asp}^2}{2}\left(1 + \lambda \frac{L}{D_{asp}}\right)$$

D_{asp}	$Q_{H20,cond}$	U_{asp}	λ	ΔH_{pompe}	$\wp_{H20,cond}$	P_{asp}
m	m³/s	m/s	–	Pa	kW	Pa
0.5	8.04	41.0	0.011	9.31×10^7	7.49×10^5	-9.30×10^7
1	8.04	10.2	0.011	2.94×10^6	2.36×10^4	-2.84×10^6
1.5	8.04	4.6	0.011	3.91×10^5	3.14×10^3	-2.90×10^5
2	8.04	2.6	0.011	9.43×10^4	759	6.64×10^4
2.5	8.04	1.6	0.011	3.18×10^4	256	6.91×10^4
3	8.04	1.14	0.011	1.35×10^3	108.5	8.75×10^4
3.5	8.04	0.84	0.011	6.82×10^3	54.8	9.42×10^4
4	8.04	0.64	0.011	4.00×10^3	32.2	9.70×10^4
4.5	8.04	0.51	0.011	2.67×10^3	21.5	9.83×10^4
5	8.04	0.41	0.011	1.99×10^3	16.0	9.90×10^4

Table 7.13. *Calculation of the head loss in the deep cold water suction pipe and of the pressure at the pump inlet as a function of the diameter of this pipe*

For a bitumen cast iron pipe, taking the roughness equal $e = 0.05$ mm, the roughness coefficient $k = e/D_{asp}$ varies between 10^{-4} for $D_{asp} = 0.5$ m and 10^{-5} for $D_{asp} = 5$ m.The linear head loss coefficient varies in the interval $0.01 < \lambda < 0.012$. Taking an average value for all cases $\lambda = 0.011$, Table 7.13 summarizes the results relating to the pumping of cold water from the bottom of the sea. The conclusion of this table is the need to have a pipe with a diameter D_{asp} greater than 3 m. This diameter must be greater than 2 m to protect against cavitation (the pressure must be greater than the vapor pressure). These are the shaded values in the last column of the table. The diameter must also be greater than 3.5 m (shaded values in the penultimate column) to limit the pump power to a value below 50 kW. Pumping deep water is therefore a delicate operation in reality.

18) Table 7.14 summarizes the main characteristics of the design of the OTEC unit studied in this exercise.

Mechanical power supplied	**Evaporator**					
	Thermal power	**Pressure**	**Water flow rate**	**Number of tubes**	**Tube length**	**Volume**
MW	**MW**	**Bar**	**m³/s**	**–**	**m**	**m³**
1	102.1	8,575	1.28	2,500	81.6	510
Condenser						**Suction**
Thermal power	**Pressure**	**Water flow**	**Number of tubes**	**Tube length**	**Volume**	D_{asp}
MW	**Bar**	**m³/s**	**–**	**m**	**m³**	**m**
100.8	6,152	8.04	13,225	10.7	352	3.5

Table 7.14. *Summary table of evaporator, condenser and deep cold water suction line sizing*

For a supplied mechanical power of 1 MW, the size of the dimensioned OTEC unit is large.

Due to the reduced difference between the warm sea temperature (32°C) and the cold sea temperature (5°C), the yield is low. The thermal powers involved are very significant. If the operating conditions of the evaporator and the condenser are different in terms of water flow, number of tubes and tube length, the volumes of the two exchangers are comparable: 510 m³ for the evaporator versus 352 m³ for the condenser.

The gigantism of the unit is reflected in the diameter of the deep-water suction pipe.

The large size of the installation leads to significant risks in terms of safety, due to the high pressures in the two exchangers and their large volumes. The toxicity of ammonia being high, the consequences of an accident can be very serious.

The thermal powers being significant, the volume of hot water captured on the surface at a temperature of 32°C is released into the surface waters of the sea at a temperature of 13°C, while the cold water pumped from the bottom of the sea is released into surface waters at a temperature of 8°C. In the long term, these discharges are likely to cool surface water. From a process point of view, this is obviously not what we are looking for because the performance of the unit will be impaired. Biologists also point out the usefulness of assessing the impacts on marine fauna and flora. Note that 7 m^3 of water released per second corresponds to approximately 1 Mm^3 per day, which is significant. Other uses of cold water, rather than discharging it directly into the sea, can be considered.

7.3. State of play of the OTEC sector

The principle of the OTEC process is to produce mechanical energy by implementing a Rankine cycle. This thermodynamic cycle uses a warm seawater source and a cold seawater source. The turbine, which recovers mechanical energy when the gas expands, drives an alternator to produce electrical energy.

The resolution of the exercise presented in section 7.2 highlights the salient points of the OTEC process:

– The small difference between the temperature of the sea surface and the temperature in the deep waters means that the efficiency of the process is low. In the exercise, the thermal power transmitted by the hot water is worth 100 times the mechanical power provided by the thermodynamic cycle.

– The sizing of the OTEC process is a compromise between the choice of the operating conditions of the cycle and the choice of the operating parameters of the heat exchangers. The evaporator and the condenser are large size exchangers. Their sizing must ensure that speeds in the exchangers are limited. They must be sufficient for the flow to be in turbulent regime and thus promote heat transfers, without being too large so as not to penalize the energy balance of the process by high head losses in the exchangers.

– The Rankine cycle uses ammonia as the working fluid. The advantage of ammonia is that it is easily available and that the vapor pressure varies significantly with temperature in the temperature range considered. This advantage in terms of energy becomes a disadvantage in terms of risks. The evaporator and condenser are large pieces of equipment operating at high pressure. The risks are significant and the implementation of security rules is complex and costly. In the event of an accident, in addition to the damage caused by the explosion, the toxicity and dangers of ammonia for humans and the environment must be considered.

– As an OTEC unit must be close to a source of warm sea water and a source of cold sea water, the potential locations for installation are limited. They are located in tropical and equatorial zones where the underwater relief allows access to the deep seabed at short distances. The volumes of water pumped into the sea are significant. Pumping deep water requires large diameter tubes. The mechanical sustainability of pipes in the extreme conditions of storms or typhoons which regularly occur in these places must be taken into account.

– Finally, the cold water pumped from the ocean floor is released to the sea surface after passing through the condenser and after being mixed with the hot water leaving the evaporator. As the flow rates are high (remember that the efficiency of the process is low), these discharges are likely to reduce surface temperatures in the vicinity of the plant in the long term and to have environmental repercussions on fauna and maritime flora.

The OTEC concept is old. G. Claude, one of the founders of the company Air Liquide, and P. Boucherot carried out the first experiments in the 1920s and 1930s. Their failure to move beyond the experimental stage can be explained by the difficulties mentioned above.

The most successful installation to date is the OTEC unit studied and built in Hawaii since 1980 under the aegis of the National Energy Laboratory of Hawaii. It has been operational since 2015 and delivers a power of 105 kW.

France, with its overseas territories, is not left out of OTEC projects. The company DNCS promotes projects in this area. Work is being carried out in Réunion which involves university laboratories on the island. With the support of the European Commission, DCNS launched the NEMO (New Energy for Martinique and Overseas) project in 2014 for a 16 MW OTEC installation. But this project was refused in April 2018 by the Martinique assembly due to its high cost, low potential production and environmental impacts.

To be complete on the OTEC sector, it is necessary to mention certain proposals to operate the OTEC installation by implementing an open Rankine cycle. This process makes it possible to overcome the constraints and risks linked to the use of ammonia as the working fluid of the Rankine cycle. The open Rankine cycle directly evaporates warm seawater that is then condensed. The vapor pressure of water at a temperature of 2°C is 740 Pa, while it is 4,230 Pa at a temperature of 30°C. Evaporation and condensation are therefore carried out at low pressure. The volumes of the pipes and equipment of the thermodynamic loop where the water is in vapor form (evaporator, condenser and turbine) must be gigantic to provide even a little energy.

7.4. Guide to sizing a heat exchanger

A heat exchanger is a device that transfers heat from a hot fluid to a cold fluid. Figure 7.5 shows the schematic of a shell and tube type heat exchanger that we discuss in a general context. The shell is the exterior casing of the device. The hot fluid flows through cylindrical tubes contained within the shell while the cold fluid flows through the shell between the tubes. The flow of the two fluids is axial along the axis of the tubes and countercurrent, that is to say that the two flows are directed in opposite directions. The transfer surface between the two fluids, denoted S, is the surface of the N cylindrical tubes contained in the casing.

The operating parameters of the exchanger are as follows:

– the mass flow rates of the cold and hot fluids, respectively, $\dot{m}_c$ and $\dot{m}_h$;

– the temperatures at the inlet of the cold and hot fluids, respectively, $T_{c,in}$ and $T_{h,in}$;

– the temperatures at the outlet of the cold and hot fluids, respectively, $T_{c,out}$ and $T_{h,out}$.

Since the hot fluid transfers heat to the cold fluid, $T_{h,in} > T_{c,out}$ while $T_{c,in} < T_{h,out}$. The four temperatures are linked by the enthalpy balance on each of the fluids, which determines the heat flow Φ exchanged. Under the ideal hypothesis that the exchanger does not suffer any losses, the heat released by the hot fluid is recovered entirely by the cold fluid.

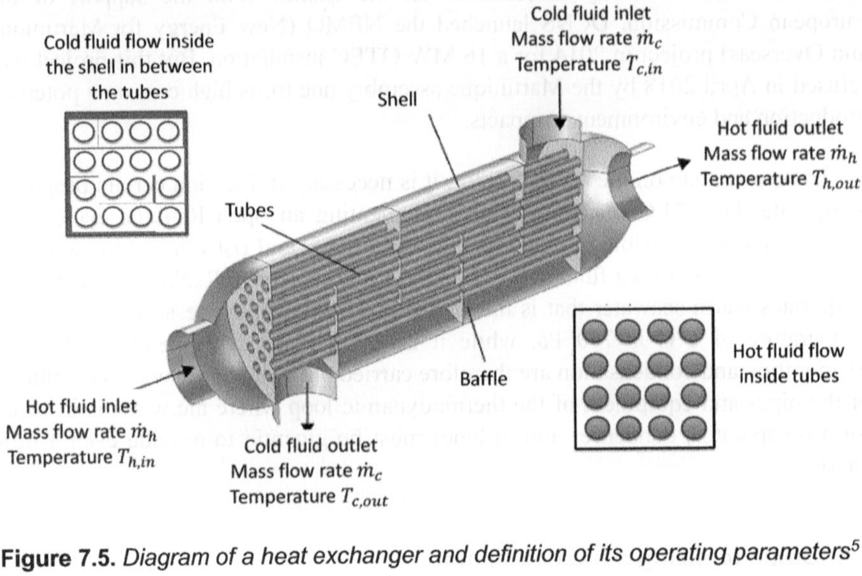

Figure 7.5. *Diagram of a heat exchanger and definition of its operating parameters*[5]

5 Source: Al Hadad, W., Schick, V., Maillet, D. (2018). Caractérisation transitoire d'un échangeur de chaleur à tubes et calandre par identification des fonctions de transfert. In *Congress of the French Thermal Society 2018*, Pau [Online]. Available at: https://hal.science/hal-02441 071/document.

In an evaporator, the cold fluid enters entirely in liquid form and exits entirely in gaseous form. The hot fluid does not undergo a phase change. The enthalpy balance is written for the two fluids:

$$\Phi = \dot{m}_h c_{p,h}\left(T_{h,in} - T_{h,out}\right)$$
$$= \dot{m}_c\left\{c_{p,c,l}\left(T_{c,evap} - T_{c,in}\right) + L_{v,c} + c_{p,c,g}\left(T_{c,out} - T_{c,evap}\right)\right\}$$

Similarly, in a condenser the hot fluid enters entirely in gas form and exits entirely in liquid form. The cold fluid does not undergo a phase change and the enthalpy balance is written as:

$$\Phi = \dot{m}_c c_{p,c}\left(T_{c,out} - T_{c,in}\right)$$
$$= \dot{m}_h\left\{c_{p,h,g}\left(T_{h,in} - T_{h,cond}\right) + L_{v,h} + c_{p,h,l}\left(T_{h,cond} - T_{h,out}\right)\right\}$$

The variables c_p designate the specific heats, whose index h or c indicates that it is the hot fluid or the cold fluid, while the index l or g indicates that it is the liquid or gaseous form.

The latent heat of the cold fluid $L_{v,c}\left(T_{c,evap}\right)$ or that of the hot fluid $L_{v,h}\left(T_{h,cond}\right)$ varies with the evaporation or condensation temperature, which are fixed by the pressure prevailing in the fluid which changes phase. If we modify the pressure, the phase change temperature is modified.

When sizing the exchanger, either the mass flow rates of the fluids are known and the temperatures of the two fluids at the outlets are deducted from the thermal power transferred. Alternatively, the outlet temperatures of the two fluids can also be fixed, in which case the necessary mass flow rates are determined from the thermal power transferred.

The power transferred Φ from the hot fluid to the cold fluid is also expressed as a function of an overall heat transfer coefficient K, the exchange surface between the fluids S (which for the exchanger shown in Figure 7.5 is the surface of the N cylindrical tubes contained in the shell) and the average logarithmic temperature difference between the two fluids, defined by[6]:

6 This expression appears when modeling a tubular exchanger. It is used for many other exchangers. When the temperature difference between the two fluids is the same at both ends of the reactor (i.e. when $T_{h,in} - T_{c,out} = T_{h,out} - T_{c,in}$), the above equation is indeterminate. However, we show by a passage to the limit that in this case $\Delta T_{mLn} = T_{h,in} - T_{c,out} = T_{h,out} - T_{c,in}$ (see the solution to question 9 of section 7.2.2).

$$\Delta T_{mLn} = \frac{(T_{h,in} - T_{c,out}) - (T_{h,out} - T_{c,in})}{Ln\left(\frac{T_{h,in} - T_{c,out}}{T_{h,out} - T_{c,in}}\right)}$$

The thermal power transferred is therefore also expressed in the form:

$$\Phi = \mathrm{K\,S}\,\Delta T_{mLn}$$

The sizing of an exchanger consists of determining its transfer surface S, knowing the overall exchange coefficient K. We deduce the dimensions of an exchanger for a given geometry. For the tube and shell exchanger shown in Figure 7.5, for example, the transfer surface S is fixed by the number of tubes N, their diameter D and their length L. The spacing between two neighboring tubes then makes it possible to determine the width of the grille.

The most delicate step in sizing an exchanger is determining the overall heat transfer coefficient K. Its unit is W/m²/K. It reflects the ability of fluids to transfer heat from the hot fluid to the cold fluid by passing through the solid that separates them. It depends on the thermo-physical properties of each fluid (density ρ, dynamic viscosity μ, specific heat c_p, thermal conductivity Λ) and the flow (speed, laminar or turbulent regime, space scale). For each of the two fluids, a transfer coefficient is determined, noted h_h for the heat transfer coefficient in the hot fluid and h_c for the heat transfer coefficient in the cold fluid, and we do the same to evaluate the transfer coefficient in the solid medium that separates the two fluids.

For the exchangers considered for OTEC processes, the solid separating the two fluids is metallic and it conducts thermal energy very well. The thickness of metal tubes (like those shown schematically in Figure 7.5) is small compared to the diameter of the tubes. In this case, the overall transfer coefficient is simply expressed as a function of the partial heat transfer coefficient in the cold fluid h_c and the heat transfer coefficient in the hot fluid h_h:

$$\frac{1}{K} = \frac{1}{h_h} + \frac{1}{h_c}$$

Table 7.15 shows the range of variation of the partial heat transfer coefficient in a fluid depending on the type of transfer, the flow and the state of the fluid.

Fluid type	Thermal transfer type	Transfer coefficient h_{fluid} (in W/m²/K)
Gas	Natural convection	$10 < h_{fluid} < 30$
Liquid	Natural convection	$100 < h_{fluid} < 900$
Gas	Forced convection	$30 < h_{fluid} < 300$
Liquid	Forced convection	$300 < h_{fluid} < 10{,}000$
Gas–liquid	Condensation	$50{,}000 < h_{fluid} < 110{,}000$
Gas–liquid	Evaporation	$3{,}000 < h_{fluid} < 50{,}000$

Table 7.15. *Range of variation of the partial heat transfer coefficient depending on the fluid and the type of transfer*

The heat transfer coefficient in a fluid varies significantly for the same fluid and the same type of flow. However, the mathematical form of the relationship linking the overall transfer coefficient to the partial heat transfer coefficients h_c and h_h attenuates the consequences of this observation. A simple algebra exercise shows that:

$$\frac{1}{2}\min\{h_h, h_c\} < K < min\{h_h, h_c\}$$

It is therefore important to know the smaller of the two transfer coefficients h_c and h_h, because it is this that sets the order of magnitude of the overall transfer coefficient K.

We also check that $K = h_c/2$ when $h_c = h_h$, while $K \cong h_c$when $h_c \ll h_h$. It is therefore useless to precisely estimate the transfer coefficient in one fluid if we know that this transfer coefficient is large compared to that in the other fluid. It is more essential to know as precisely as possible the value of the transfer coefficient in the fluid where it is lowest.

An analogy with automobile traffic on a highway illustrates the notion of partial transfer coefficient. The Saint–Arnoult toll on the A10 motorway in France is known to generate large traffic jams on busy days. Suppose that the Saint–Arnoult toll allows 40 vehicles to pass per minute. This figure is the partial transfer coefficient in the toll. If the automobile flow on the highway is 100 vehicles per minute, the overall flow on a section including the toll will be 40 vehicles per day. It is the Saint–Arnoult toll which limits traffic. It is necessary to increase its transfer coefficient to reduce traffic jams. Conversely, if the automobile flow on the highway is five vehicles per minute, it fixes the transfer through the toll. The Saint–Arnoult toll is not limiting in this situation.

As elementary as they are, the principles of sizing a heat exchanger presented in this section are sufficient to achieve relevant dimensions in order of magnitude of the evaporator and the condenser for the OTEC process studied in the exercise in section 7.2. They make it possible to not neglect these dimensions and to draw up a relevant inventory of the process as a whole.

8

Wave Energy

8.1. Surface gravity wave theory

8.1.1. *General*

The exercise proposed in section 8.2 studies the recovery of wave energy using a flap that oscillates under the action of gravity waves. The waves apply a torque to the flap and cause it to perform a sinusoidal rotational movement around an axis. Controlling the oscillation movement of the flap by a PTO (Power Take Off) system makes it possible to recover the energy of the waves in the form of mechanical power of the flap.

Formally, the resolution of the exercise relates to the mechanics of non-deformable solids. This first section, which summarizes knowledge about wave dynamics[1], contextualizes the exercise. It also makes it possible to differentiate the wave energy recovery systems, which are presented in section 8.3.

A wave first appears as the deformation of the ocean surface that rises above and falls below resting sea level. These displacements of the water plane form regular patterns on the ocean surface, as we see in Figure 8.1 where the crests of parallel waves are spaced by a length λ called wavelength. If we fix a point on the surface of the ocean, we observe the temporal oscillation with period T of its altitude in the vertical direction z (oriented upwards in the direction opposite to the acceleration of gravity). We also note the movement of the wave crest in a direction perpendicular (x direction) to the wave crest (y direction). The speed of propagation $c = \lambda/T$ is called phase speed. The more complex patterns observed on the sea surface can

1 The late Jean-Pierre Germain could recognize in this section numerous elements which, when they are not borrowings, are inspired by his course taught in Grenoble and his booklets.

be decomposed into a superposition of waves of different periods, wavelength, amplitude and direction of propagation. Figure 8.1 helps to understand why they are called surface gravity waves. Gravity plays a central role in the dynamics of vertical shifts in surface position. The geometry of the deformations moves in space with time, hence the term wave assigned to them.

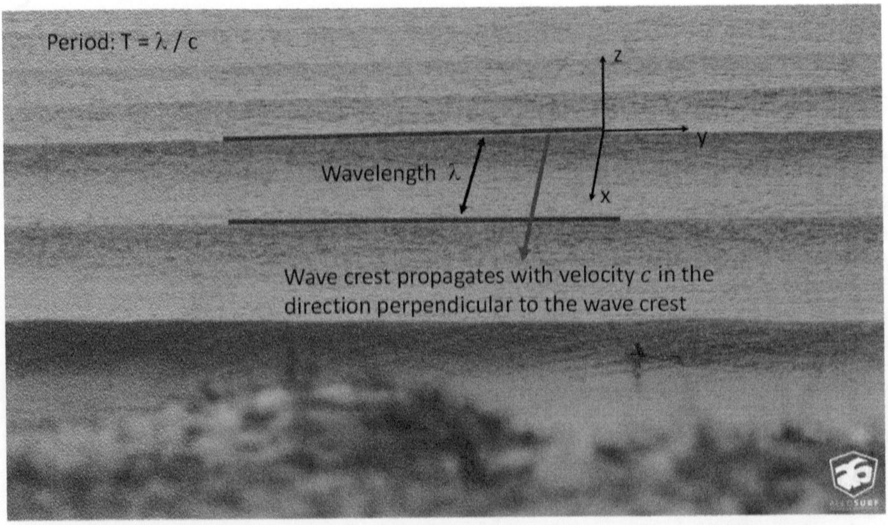

Figure 8.1. *Observation of a gravity wave on the surface of the ocean[2]. For a color version of this figure, see www.iste.co.uk/mory/mechanical.zip*

Surface gravity waves are movements that propagate in a water layer initially at rest. The effect of viscosity is negligible (large-scale movements); water is incompressible. By virtue of Kelvin's theorem, gravity waves are irrotational flow because the curl of the velocity is zero ($\overrightarrow{Rot}(\vec{u}) = \vec{0}$). The velocity field can be described using a velocity potential function ϕ (see section 1.10), which is a harmonic function. This is explained by:

$$\vec{u}(x, y, z, t) = \nabla\phi(x, y, z, t)$$

$$\Delta\phi = 0$$

We are looking for a particular wave-type solution to these equations.

For a traveling wave propagating in the direction $x > 0$, the velocity potential function depends on the following variables $\phi(x - ct, z)$. Indeed, any movement

2 Available at: https://www.allosurf.net/meteosurf-faq.php.

observed at x_0 at time t is reproduced identically at $x_1 > x_0$ at time $t' = t + (x_1 - x_0)/c >$, because:

$$\phi(z, x_1 - ct') = \phi(z, x_0 - ct)$$

Analogously for a wave propagating in the direction $x < 0$, the velocity potential function takes the form $\phi(z, x + ct)$.

Any wave solution can be developed in a Fourier series whose eigenmodes are of the form:

$$\phi(z, x - ct) = \Phi(z)\sin(kx - \omega t + \varphi)$$

This expression shows the angular frequency ω, linked to the period:

$$\omega = \frac{2\pi}{T}$$

The wave number k is linked to the wavelength:

$$k = \frac{2\pi}{\lambda}$$

The phase speed of the wave can therefore be written in two forms:

$$c = \frac{\omega}{k} = \frac{\lambda}{T}$$

φ designates the phase of the wave, which without losing generality can be taken equal to $\varphi = 0$.

Each eigenmode is a harmonic function at any time and at any point:

$$\Delta\phi = \left\{-k^2\Phi(z) + \frac{d^2\Phi(z)}{dz^2}\right\}\sin(kx - \omega t) = 0$$

The function $\Phi(z)$ is therefore of the form:

$$\Phi(z) = Ae^{kz} + Be^{-kz}$$

The integration constants A and B depend on the domain in which the wave propagates.

8.1.2. *Kinematics of linear surface waves*

For a wave propagating in a water layer whose depth is constant when the fluid is at rest (Figure 8.2, we note it h), we naturally fix the altitude $z = 0$ on the bottom (assumed here horizontal, the acceleration of the gravity being perpendicular to it). The speed must verify the boundary condition:

$$u_z(z = 0, x - ct) = \frac{d\Phi(z = 0)}{dz}\sin(kx - \omega t) = 0$$

Therefore:

$$\phi(z, x - ct) = A ch(kz)\sin(kx - \omega t) \quad [8.1]$$

For a wave propagating in a water layer of infinite depth (Figure 8.3), the altitude $z = 0$ is chosen at the sea surface at rest. Since the speed must remain finite when $z \to -\infty$, the solution is written as:

$$\phi(z, x - ct) = Ae^{kz}\sin(kx - \omega t) \quad [8.2]$$

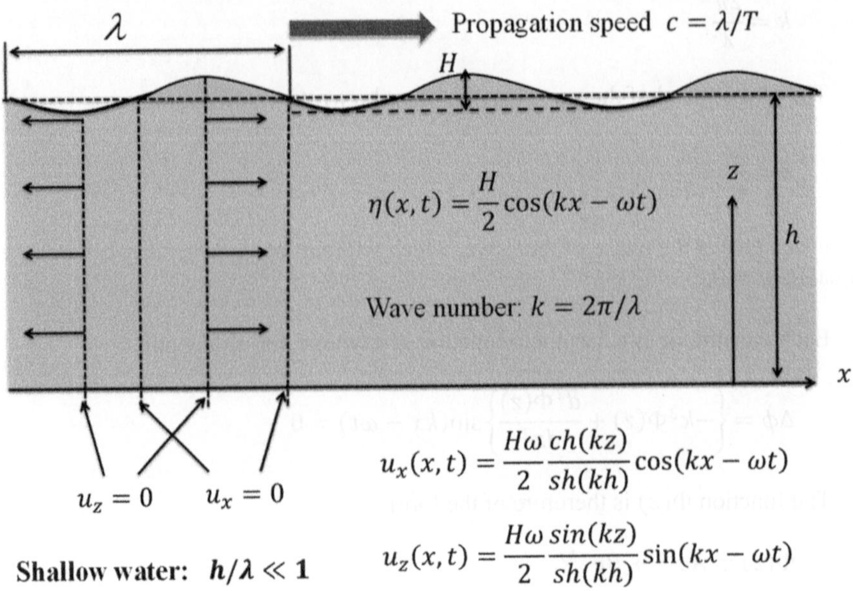

Figure 8.2. *Kinematics of a surface gravity wave in a water layer of constant finite depth*

The next step consists of relating the velocity field in the water layer to the deformation of the free surface caused by the wave. This link is made by writing the kinematic condition at the free surface: in the frame moving at phase speed c, the deformation of the free surface and the velocity field are stationary. The speed must be tangent to the free surface, which is written as:

$$\frac{\partial \eta}{\partial t} + u_x \frac{\partial \eta}{\partial x} = u_z(x, z = h + \eta, t)$$

where $\eta(x,t)$ is the vertical displacement of the surface generated by the wave relative to its altitude at rest.

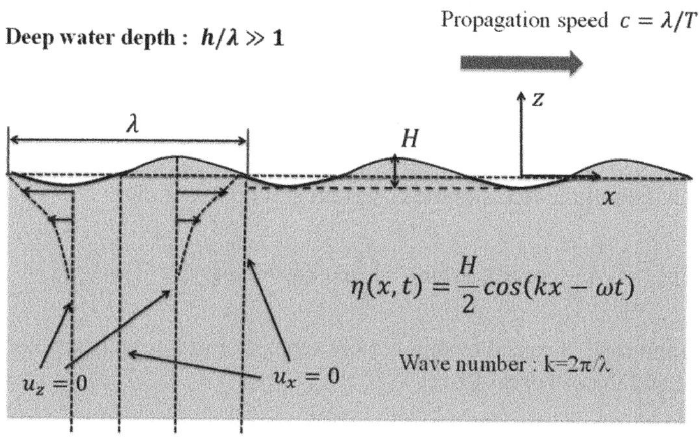

$$u_z(x,t) = \frac{H\omega}{2} exp(kz) sin(kx - \omega t) \qquad u_x(x,t) = \frac{H\omega}{2} exp(kz) cos(kx - \omega t)$$

Figure 8.3. *Kinematics of a surface gravity wave in a water layer of infinite depth*

The linear wave solution solves the equation by linearization with respect to the disturbance.

For a wave propagating in a water layer of finite depth, we therefore write using equation [8.1]:

$$\frac{\partial \eta}{\partial t} \cong u_z(x, z = h, t) = \frac{d\phi(z = h)}{dz} = Aksh(kh)\sin(kx - \omega t)$$

Its temporal integration determines the deformation of the free surface:

$$\eta(x-ct) = A\frac{k}{\omega}sh(kh)\cos(kx-\omega t) = \frac{H}{2}\cos(kx-\omega t)$$

We deduce from this equation the relationship between the crest to trough height difference H of the wave and the constant A in the expression of the velocity potential:

$$\frac{H}{2} = A\frac{k}{\omega}sh(kh)$$

Analogously, for a wave propagating in a water layer of infinite depth, we use equation [8.2] to deduce:

$$\frac{\partial\eta}{\partial t} \cong u_z(x,z=0,t) = \frac{d\phi(z=0)}{dz} = Ak\sin(kx-\omega t)$$

and the deformation of the free surface is, by integration over time:

$$\eta(x-ct) = A\frac{k}{\omega}\cos(kx-\omega t) = \frac{H}{2}\cos(kx-\omega t)$$

In this second case, the relationship between the crest to trough height difference H of the wave and the constant A is:

$$\frac{H}{2} = A\frac{k}{\omega}$$

The velocity components are ultimately deduced by deriving the velocity potential:

$$u_x(z,x,t) = \frac{\partial\phi}{\partial x}$$

$$u_z(z,x,t) = \frac{\partial\phi}{\partial z}$$

For both cases, the expressions for the velocity components $u_x(z,x,t)$ and $u_z(z,x,t)$ are written in Figures 8.2 and 8.3.

The remarkable properties of the velocity field, useful for understanding the operation of wave energy converters, are that:

– Below the wave crests ($\eta = H/2$) the speed component u_x reaches its maximum value and it is oriented in the same direction as the direction of propagation of the wave. The speed component u_z is 0.

– Below the wave troughs ($\eta = -H/2$), the speed component u_x reaches its maximum negative value and it is oriented in the direction opposite to the direction of propagation of the wave. The speed component u_z is 0.

– Below positions where $\eta = 0$, the speed component u_x is 0 while the speed component u_z is maximum, negative or positive.

– The plot in Figure 8.2 corresponds to the case of a wave propagating in shallow depth ($kh = 2\pi h/\lambda \ll 1$). In this particular case, the component u_x does not vary along a vertical line.

A second boundary condition at the surface, the dynamic condition, fixes a relationship between period and wavelength for a given depth of the water layer at rest.

The flow being irrotational, it verifies Bernoulli's theorem for an incompressible fluid[3]:

$$\rho\frac{\partial\phi}{\partial t} + \rho\frac{u_x^2 + u_z^2}{2} + p + \rho g z = \text{constant}$$

This condition must be verified at the free surface where atmospheric pressure prevails. Hence, for the finite depth configuration:

$$\rho\frac{\partial\phi}{\partial t}(x, z = h + \eta, t) + \rho\frac{u_x^2 + u_z^2}{2}(x, z = h + \eta, t) + P_{atm} + \rho g\eta$$
$$= \text{constant}$$

The constant is necessarily equal to atmospheric pressure.

3 Although it applies to unsteady flows, this Bernoulli theorem is less general than the one presented in section 1.2, because it requires that the flow be irrotational. For an irrotational flow the head is constant throughout the domain. For the theorem presented in section 1.2, the head is constant on a current line but its value can be different on two distinct current lines.

Linearizing this equation with respect to the disturbance amounts to neglecting the kinetic energy and approximating $h + \eta \cong h$. From where:

$$\rho \frac{\partial \phi}{\partial t}(x, z = h, t) + \rho g \eta = 0$$

By introducing into this equation the solutions of ϕ and η (equation [8.1] and Figure 8.2), we obtain:

$$-\rho \omega A ch(kh) \cos(kx - \omega t) + \rho g \frac{A k sh(kh)}{\omega} \cos(kx - \omega t) = 0$$

from which we deduce the dispersion relation:

$$\omega^2 = g k th(kh) \quad [8.3]$$

For a given depth, the angular frequency ω and the wave number k are therefore linked, which also means that the period T and the wavelength λ are also linked. The longer the period, the longer the wavelength.

The dispersion relation makes it possible to calculate the phase speed of the wave as a function of the period T or wavelength λ alone. Two special cases deserve to be known:

– that of long waves , when $kh \ll 1$: then $\omega^2 \cong g k^2 h$. We then deduce:

$$c = \frac{\omega}{k} = \frac{\lambda}{T} = \sqrt{gh}$$

– that of short waves , when $kh \gg 1$: then $\omega^2 \cong gk$. We then deduce:

$$c = \frac{\omega}{k} = \frac{\lambda}{T} = \frac{gT}{2\pi} = \sqrt{g\lambda/2\pi}$$

8.1.3. *Wave energy and energy flow*

For a wave propagating in the direction $x > 0$, the instantaneous flow of energy through a section of a vertical plane perpendicular to the direction of propagation of the wave, of width 1 m (in the y direction) and of water depth $h + \eta(x, t)$, is equal to the flow of head through this plane (see section 1.7). This is written as:

$$\Im(x, t) = \int_0^{h+\eta} \left\{ \rho \frac{u_x^2 + u_z^2}{2} + p + \rho g z \right\} u_x dz$$

Using the Bernoulli relation for irrotational flow, introduced in section 8.1.2, and expressing the component u_x using the velocity potential function, this expression takes the form:

$$\Im(x,t) = -\int_0^{h+\eta} \rho \frac{\partial\phi}{\partial t}\frac{\partial\phi}{\partial x} dz$$

The energy flow per unit wave crest width depends on the velocity potential alone $\phi(x,z,t)$. By replacing the velocity potential function by its expression (equation [8.1]) for a linear wave propagating in a layer of finite thickness h, we obtain:

$$\Im(x,t) = \rho \frac{H^2}{4}\frac{\omega^3}{k}\frac{1}{sh^2(kh)}\cos^2(kx-\omega t)\int_0^{h+\eta} ch^2(kz)dz$$

The square of the cosine means that the wave carries in the direction of propagation of the wave a non-zero average flux of energy per unit width of the wave crest:

$$\begin{aligned}\bar{\Im}_{wave} &\cong \rho \frac{H^2}{4}\frac{\omega^3}{k}\frac{1}{sh^2(kh)}\frac{1}{2}\int_0^h ch^2(kz)dz \\ &= \rho \frac{H^2}{16}\frac{\omega^3}{k^2}\frac{1}{sh^2(kh)}\left\{kh + \frac{sh(2kh)}{2}\right\}\end{aligned} \qquad [8.4]$$

The calculation of the integral in this expression neglects η in front of h.

The instantaneous energy contained in the water layer per unit surface of the horizontal plane is the sum of the kinetic energy and the potential energy:

$$\begin{aligned}E(x,t) &= \int_0^{h+\eta}\left\{\rho\frac{u_x^2+u_z^2}{2} + \rho g z\right\}dz \\ &= \frac{1}{2}\rho\int_0^{h+\eta}(u_x^2+u_z^2)dz + \rho g\frac{(h+\eta)^2}{2} \\ &\cong \frac{1}{2}\rho\int_0^{h}(u_x^2+u_z^2)dz + \rho g\frac{h^2+2h\eta+\eta^2}{2}\end{aligned}$$

By replacing the velocity components and the displacement of the free surface by their expressions for linear waves (Figure 8.2), we deduce:

$$E(x,t) \cong \rho g \frac{h^2 + hH\cos(kx-\omega t) + \frac{H^2}{4}\cos^2(kx-\omega t)}{2}$$
$$+ \rho \frac{H^2\omega^2}{8sh^2(kh)}\cos^2(kx-\omega t)\int_0^h ch^2(kz)\mathrm{d}z$$
$$+ \rho \frac{H^2\omega^2}{8sh^2(kh)}\sin^2(kx-\omega t)\int_0^h sh^2(kz)dz$$

Its average value over time at any point on the plane is:

$$\bar{E} \cong \rho g \frac{h^2}{2} + \rho g \frac{H^2}{16} + \rho \frac{H^2\omega^2}{16sh^2(kh)}\int_0^h \{ch^2(kz) + sh^2(kz)\}dz$$
$$= \rho g \frac{h^2}{2} + \rho g \frac{H^2}{16} + \rho \frac{H^2\omega^2}{16sh^2(kh)}\int_0^h ch(2kz)dz$$
$$= \rho g \frac{h^2}{2} + \rho g \frac{H^2}{8}$$

This expression is obtained by calculating the integral in the vertical direction and using the dispersion relation [8.3]. The term $\rho g h^2/2$ is the potential energy of the water layer at rest per unit surface of the horizontal plane. By omitting it to retain only the term linked to the gravity wave, we evaluate the average energy of the wave integrated over the water layer depth per unit surface of the horizontal plane:

$$\bar{E}_{wave} = \rho g \frac{H^2}{8} \quad [8.5]$$

The remarkable property is that the average energy of the gravity wave varies with the square of the crest-to-trough wave height H.

By introducing the group speed c_g, whose expression is deduced from the dispersion relation:

$$c_g = \frac{\partial \omega}{\partial k} = \frac{g}{2\omega}\frac{1}{ch^2(kh)}\left\{kh + \frac{sh(2kh)}{2}\right\} \quad [8.6]$$

we verify, by bringing together the expressions [8.4], [8.5] and [8.6], that the average wave energy flux is written as:

$$\overline{\Im}_{wave} = \bar{E}_{wave} c_g = \rho g \frac{H^2}{8} c_g$$

The phase speed $c = \omega/k$ is the speed of movement of the wave crests while the group speed $c_g = \partial\omega/\partial k$ is the speed of transport of wave energy.

In the limiting case of long waves ($kh \ll 1$):

$$c_g = c = \sqrt{gh}$$

For the limiting case of short waves ($kh \gg 1$):

$$c_g = \frac{c}{2} = \frac{g}{2\omega} = \frac{1}{2}\sqrt{\frac{g}{k}}$$

The energy flux $\overline{\Im}_{wave}$ is the wave power per unit width of a wave crest. Its unit is W/m. Maps of the distribution of the average annual wave power on the ocean surface are easily found on the Internet[4]. Far from the shores, these are short waves whose period varies between around 10 s and a few tens of seconds. By virtue of the relationship linking the energy flow $\overline{\Im}_{wave}$ to the wave energy $\bar{E}_{wave}$, this map gives an assessment of the average crest-to-trough height of waves at the sea surface. The energy potential is considerable, but we can only envisage recovering a small fraction of it near the coasts where wave energy converters can be installed and from where the energy can then be repatriated to land.

8.1.4. *Reflection of a wave by a vertical wall*

A traveling wave is reflected by a vertical wall in such a way as to ensure that the velocity component perpendicular to the wall is canceled at any time on the wall. The solution is easy to write when the wave propagates in a water layer of constant depth. It is enough to add the solution of the incident wave propagating in the direction $x > 0$ given in Figure 8.2 to the solution of an incident wave of the same period propagating in the direction $x < 0$. The amplitude of the reflected wave is equal to that of the incident wave. The solution is explained in Figure 8.4. We check that $u_x(x = 0, t) = 0$ on the wall (in $x = 0$).

4 The keywords "Average Annual Wave power" enable visualizing the distribution map of the average annual wave power on the ocean surface, taken from Cruz (2007).

We notice that the addition of the cosine functions means that the superposition of the incident wave on the reflected wave is a standing wave. The crests of the wave (positions where the vertical displacement of the free surface is maximum) do not move in the x direction. On the vertical lines $x = 0$, $x = -\lambda/2$ and $x = -\lambda$, the position of the free surface oscillates with a peak-to-trough amplitude equal to $2H$ (double the height H of the incident wave), while on the vertical lines $x = -\lambda/4$ and $-3\lambda/4$ the position of the surface free does not move, remaining equal to $z = 0$.

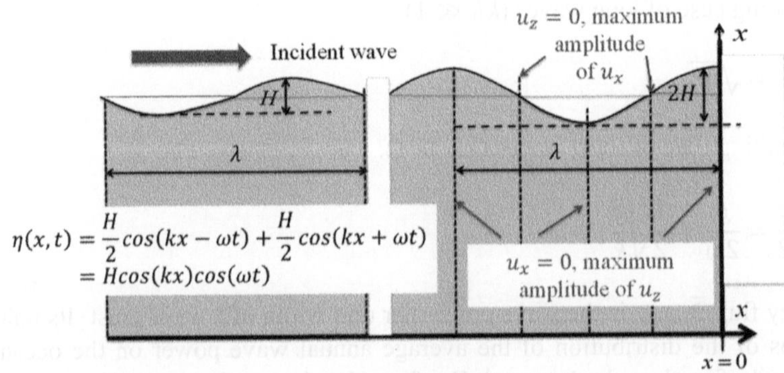

$$u_x(x,t) = \frac{H\omega}{2}\frac{ch(kz)}{sh(kh)}cos(kx - \omega t) - \frac{H\omega}{2}\frac{ch(kz)}{sh(kh)}cos(kx + \omega t) = H\omega\frac{ch(kz)}{sh(kh)}sin(kx)sin(\omega t)$$

$$u_z(x,t) = \frac{H\omega}{2}\frac{sh(kz)}{sh(kh)}sin(kx - \omega t) - \frac{H\omega}{2}\frac{sh(kz)}{sh(kh)}sin(kx + \omega t) = -H\omega\frac{sh(kz)}{sh(kh)}cos(kx)sin(\omega t)$$

Figure 8.4. *Reflection of a traveling wave by a vertical wall*

8.1.5. *A laboratory experiment achieving total recovery of the wave energy flow*

Figure 8.5 shows the schematic of a laboratory experiment in which an oscillating piston generates a progressive wave at one end of a channel. The piston wave generator is a vertical plate whose electromechanical system activates the oscillatory movement $X(0,t)$ around the position $x = 0$. If the period T is long enough ($T\sqrt{gh} \gg 1$), the velocity component $u_x(x,z,t)$ of the progressive wave generated varies little with the vertical coordinate z. On the plate, the flow velocity component is equal to the movement speed of the plate:

$$\frac{dX}{dt}(x = 0, t) = u_x(x = 0, z, t)$$

This relationship makes it possible to relate the amplitude of displacement of the plate to the height H of the generated traveling wave and to the wave propagation conditions, using the results shown in Figure 8.2. The characteristics of the wave produced are shown in Figure 8.5.

It is possible to fully recover the energy flow of the emitted wave by using a second piston, placed at the other end of the channel ($x = L$). If the displacement of the energy recovering piston corresponds to the wave velocity field at the other end of the channel $x = L$, namely:

$$\frac{dX}{dt}(x = L, t) = u_x(x = L, z, t)$$

the wave is not reflected by the energy recovering piston. The energy flow is then entirely recovered by the receiving piston. To carry out this experiment, it is enough to have an electromechanical system (PTO) capable of adjusting the movement of the energy recovering piston in amplitude and phase to the velocity field of the incoming wave.

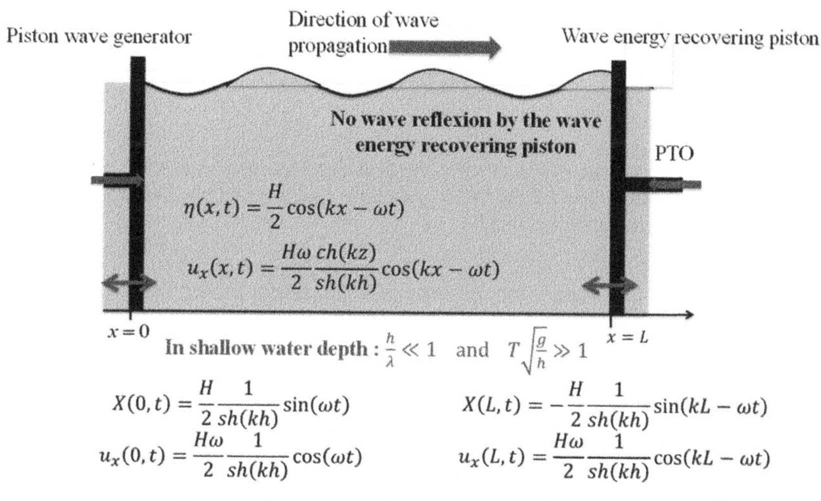

Figure 8.5. *Experiment of generation and recovery of wave energy by oscillating plane pistons in a wave channel. For a color version of this figure, see www.iste.co.uk/mory/mechanical.zip*

8.1.6. *Recovery of wave energy by an oscillating flap or by a heaving buoy*

Figures 8.6 and 8.7 schematically show two wave energy recovery systems.

The first, which is studied in the exercise proposed in section 8.2, uses a flap which oscillates under the action of an incident wave by making a rotational movement $\theta(t)$ around an axis. When the water layer is at rest, the flap is in equilibrium when its position is vertical ($\theta = 0$). When a wave propagates, it applies forces to the flap, which make it oscillate. The wave energy is converted into the kinetic energy of the oscillating panel. The latter is recovered by an electromechanical system which applies a resistant torque to the flap in order to control its movement.

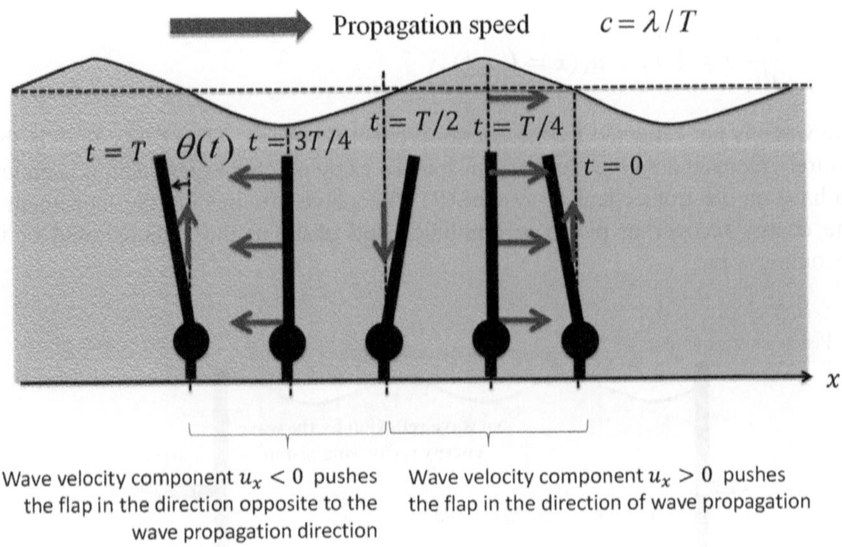

Figure 8.6. *Oscillatory movements of a flap under the action of an incident progressive wave. For a color version of this figure, see www.iste.co.uk/mory/mechanical.zip*

The second system is a buoy on the water surface which oscillates vertically in the wave field. The buoy is connected by a rod to an electromechanical system placed at the bottom which applies a resisting force and thus recovers wave energy. The movement of the float $Z(t)$ is vertical.

Rather than schematizing the advancement of the wave in relation to the flap or the float, we represent in Figures 8.6 and 8.7 the recovery device at different positions under the wave which correspond to successive moments during a period. The oscillating flap is pushed by the horizontal velocity component u_x while the float moves under the action of the vertical velocity component u_z. The flow speed is represented by the red arrows, reproducing the kinematics of flow under a wave described in Figure 8.2. Since the float is on the surface of the water, it is intuitive to understand its movement under the effect of the vertical component of speed under the wave. For the oscillating panel, it is the sign of the horizontal u_x velocity component that allows us to understand the direction of rotation of the flap according to the moment.

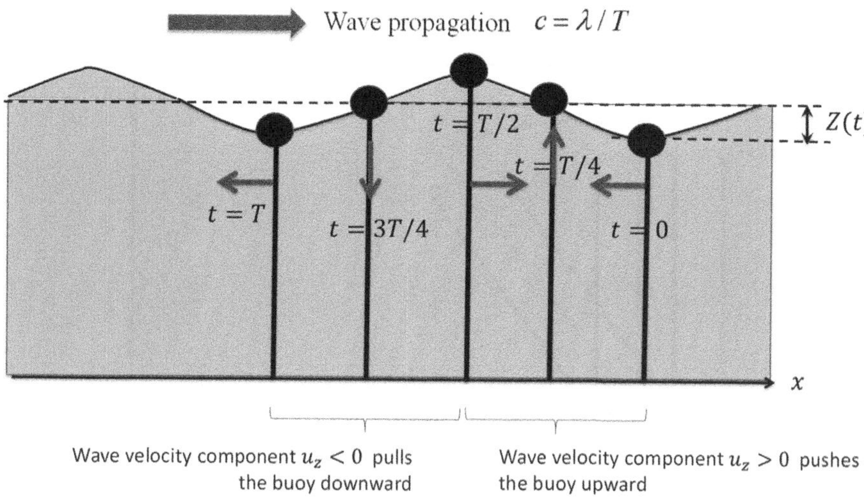

Figure 8.7. *Oscillatory movements of a float under the action of an incident progressive wave. For a color version of this figure, see www.iste.co.uk/mory/mechanical.zip*

8.2. Exercise: sizing an oscillating flap wave energy converter

We consider a flap that oscillates in order to recover gravity wave energy[5]. It is a panel of width L and height h_v immersed in water (Figure 8.8). In this exercise, we assume that the plane of the panel is perpendicular to the direction of wave propagation. The waves are assumed to be regular, with wave height H and wave period T. The flap is completely submerged. The water depth is $h > h_v$.

5 This exercise is inspired by a research work by Mory et al. (2015).

In this exercise, we quantify the wave energy which is recovered by the flap in the form of kinetic energy of its oscillating motion. The rotational movement $\theta(t)$ of the panel is described by the angular momentum theorem applied to the flap:

$$I_y\ddot{\theta}(t) = \tau_{waves}(t) + \tau_{rad}(t) + \tau_{PTO}(t) - g(\rho V - M)d_1\theta(t) \qquad [8.7]$$

I_y is the moment of inertia of the panel.

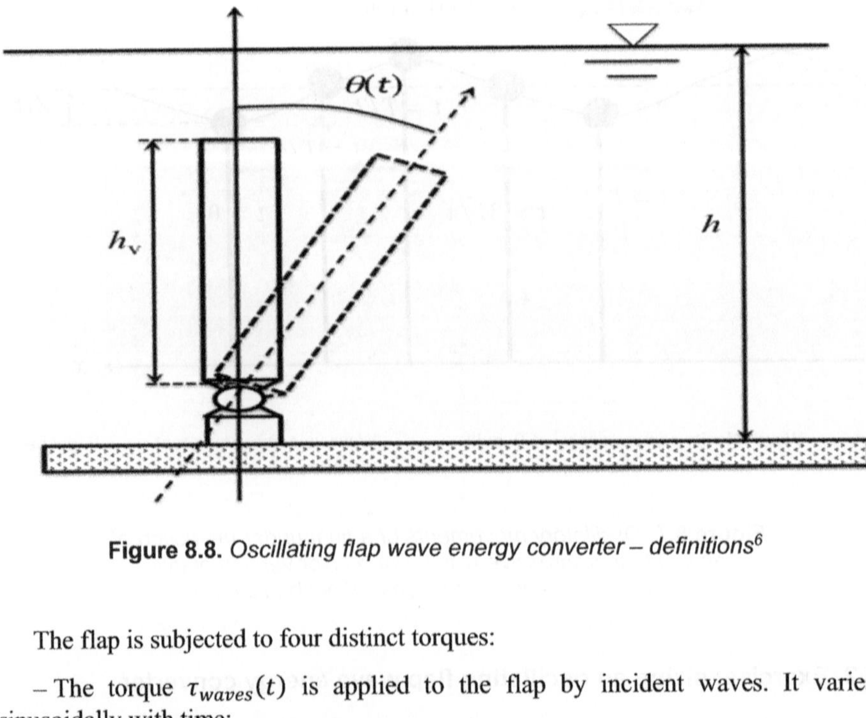

Figure 8.8. *Oscillating flap wave energy converter – definitions*[6]

The flap is subjected to four distinct torques:

– The torque $\tau_{waves}(t)$ is applied to the flap by incident waves. It varies sinusoidally with time:

$$\tau_{waves}(t) = \tau_{inc}\sin(\omega t) \quad \text{with} \quad \tau_{inc} = A\rho g H h^2 L \qquad [8.8]$$

The dimensionless constant A is evaluated using a numerical wave propagation model which takes into account the propagation of incident waves, their reflection and their diffraction by the flap. $\omega = 2\pi/T$ is the angular frequency of the waves.

– When the panel is subjected to a regular wave, it carries out a sinusoidal oscillation movement $\theta(t) = \theta_0\sin(\omega t + \varphi)$, of amplitude θ_0 and phase shifted by

6 Ibid.

an angle φ relative to the incident wave. The movement of the flap in return emits waves (radiation), associated with a torque applied to the panel. Its form is written as:

$$\tau_{rad}(t) = -C_A\dot{\theta}(t) \quad \text{with} \quad C_A = B\rho g h^3 L\sqrt{h/g} \qquad [8.9]$$

A numerical wave model determines the value of the dimensionless constant B[7].

– The recovery of wave energy is carried out using a mechanical converter (PTO) designed to apply a resisting torque to the flap:

$$\tau_{PTO}(t) = -B_{PTO}\dot{\theta}(t) - K_{PTO}\theta(t) \qquad [8.10]$$

– We recognize in the fourth term the torque linked to the weight and the Archimedean thrust which apply to the panel. M designates the mass of the panel, V its volume and d_1 is the distance from the center of gravity of the panel to its axis of rotation. The panel is a closed box filled with air. At rest, gravity keeps the panel in equilibrium in a vertical plane ($\theta = 0$).

1) Express the mechanical power recovered by the converter and calculate the energy recovered during a time interval of duration t. Comment on the result, considering in particular the case where the duration t is an integer number of wave periods $T = 2\pi/\omega$. What are the respective roles played by the coefficients B_{PTO} and K_{PTO}?

2) Introduce into the equation of angular momentum [8.7] the torque expressions introduced in the statement. Identify a resonance phenomenon and express the resonance angular frequency.

3) Introduce into the differential equation found in question 2 the flap angle variation $\theta(t) = \theta_0\sin(\omega t + \varphi)$. Specify the solution by expressing the conditions that must be verified for the phase shift angle φ and the amplitude of oscillation θ_0.

7 The relationships which respectively link τ_{inc} to the dimensionless constant A and C_A to the dimensionless constant B involve the density of water ρ, the acceleration of gravity g, the thickness of the water layer h, the width of the panel L and the wave height H. The dimensional relationships are deduced from the linear wave models. We do not demonstrate them. It is still useful to note that τ_{inc} depends on the wave height H, since it is the torque applied by the incident wave on the flap. On the other hand, C_A does not depend on H because C_A is used to evaluate the torque applied by the waves emitted by the movement of the flap, hence the dependence of the torque $\tau_{rad}(t)$ on the angular speed of rotation of the panel $\dot{\theta}(t)$.

4) Express the average power $\wp_{average}$ recovered during a wave period. Determine the conditions that optimize recovery. Determine the corresponding values of the parameters $B_{PTO,opt}$ and $K_{PTO,opt}$.

5) For conditions that maximize energy recovery, calculate the maximum average power recovered $\wp_{average,max}$. Propose an expression for estimating the efficiency r_{flap} of wave energy recovery. The water height is $h = 8$ m and the wave period $T = 14$ s. It will be assumed that the waves are long waves. Check the validity of this hypothesis.

6) For conditions that optimize recovery, determine the angle amplitude θ_0 of the flap oscillation. Comment on the result with regard to the possibility of carrying out active control of the wave energy system.

7) By introducing the variables x and y defined by:

$$\omega_R^2 = \frac{K_{PTO} + g(\rho V - M)d_1}{I_y} = \omega^2 x$$

$$B_{PTO} = B_{PTO,opt}y$$

express as a function of x and y the ratio $\wp_{average}(B_{PTO}, K_{PTO})/\wp_{average,max}$ and the angle amplitude $\theta_O(B_{PTO}, K_{PTO})$ of oscillation of the panel.

The characteristics of the panel are $I_y = 4 \times 10^6$ kg. m^2 and $L = 20$ m. The water height is $h = 8$ m and the wave period $T = 14$ s. For the three cases considered, insert the numerical results in Table 8.1.

H	***A***	***B***	θ_O	$\wp_{average,max}$	r_{flap}
m	**–**	**–**	**°**	**kW/m**	**–**
1	0.2	0.1			
1	0.4	0.2			
2	0.4	0.2			

Table 8.1. *Characteristics of wave energy recovery by an oscillating flap*

Superimpose on the same graph the isocontours of the functions $\theta_O(B_{PTO}, K_{PTO})$ and $\wp_{average}$ $(B_{PTO}, K_{PTO})/\wp_{average,max}$. Comment on the figure.

8.2.1. *Solution*

1) Wave energy is recovered by the PTO system which applies the torque to the flap:

$$\tau_{PTO}(t) = -B_{PTO}\dot{\theta}(t) - K_{PTO}\theta(t)$$

The mechanical power recovered by the PTO is:

$$\wp(t) = \tau_{PTO}(t)\dot{\theta}(t) = -B_{PTO}\dot{\theta}^2(t) - K_{PTO}\theta(t)\dot{\theta}(t)$$

For positive B_{PTO}, the first term to the right of the equality is negative. This means that the PTO system recovers energy by applying a resistive force to the panel.

The energy recovered during a time interval of duration t is:

$$\begin{aligned} E(t) &= -\int_0^t \wp(t')dt' = B_{PTO}\int_0^t \dot{\theta}^2(t')dt' + K_{PTO}\int_0^t \theta(t')\dot{\theta}(t')dt' \\ &= B_{PTO}\int_0^t \dot{\theta}^2(t')dt' + K_{PTO}\left\{\frac{\theta^2(t)}{2} - \frac{\theta^2(0)}{2}\right\} \end{aligned}$$

Energy recovery is linked to the coefficient B_{PTO}. The term related to K_{PTO} is zero if the duration t is an integer number of wave periods.

We can expect that there will be a value B_{PTO} maximizing the energy recovered. If $B_{PTO} = 0$, no energy is recovered due to the absence of a resisting torque. If B_{PTO} is very large, the resisting torque applied by the PTO is so high that the angular oscillation velocities will be very small and the energy recovery will be reduced. The compromise is sought for an intermediate value of B_{PTO}.

2) By introducing the different torque expressions into the angular momentum equation, we obtain after grouping the terms according to the orders of derivation of the flap angle $\theta(t)$:

$$I_y\ddot{\theta}(t) + \{C_A + B_{PTO}\}\dot{\theta}(t) + \{K_{PTO} + g(\rho V - M)d_1\}\theta(t) = \tau_{inc}\sin(\omega t)$$

This is a second order linear differential equation with constant coefficients which is forced by a sinusoidal term applied by the incident wave.

The resonance angular frequency is:

$$\omega_R = \sqrt{\frac{K_{PTO} + g(\rho V - M)d_1}{I_y}}$$

If we neglect the friction term (case where $C_A + B_{PTO} = 0$), the differential equation reduces to:

$$\ddot{\theta}(t) + \omega_R^2\theta(t) = \frac{\tau_{inc}}{I_y}\sin(\omega t)$$

Without forcing ($\tau_{inc} = 0$), the system oscillates naturally with the angular frequency ω_R.

If sinusoidal forcing with an angular frequency ω is applied on the device, two scenarios arise:

– when $\omega \neq \omega_R$ the solution of the differential equation is $\theta(t) = E\cos(\omega t) + F\sin(\omega t)$. The system responds with the angular frequency of the forcing. The constants E and F are determined by identification after having introduced the solution into the differential equation;

– when $\omega = \omega_R$, the solution is in the form $\theta(t) = E\ t\ cos(\omega t)$. Since:

$$\dot{\theta}(t) = E\cos(\omega t) - E\ t\ \omega\sin(\omega t)$$

$$\ddot{\theta}(t) = -2\ E\ \omega\sin(\omega t) - E\ t\ \omega^2\cos(\omega t)$$

By replacing in the differential equation:

$$\ddot{\theta}(t) + \omega_R^2\theta(t) = -2E\omega\sin(\omega t) - Et\omega^2\cos(\omega t) + \omega_R^2 Et\cos(\omega t) = \frac{\tau_{inc}}{I_y}\sin(\omega t)$$

we deduce the value:

$$E = -\frac{\tau_{inc}}{2\omega I_y}$$

We recognize in the case $\omega = \omega_R$ a classic case of resonance: the amplitude of the oscillation movement increases linearly over time. The friction terms (when $C_A + B_{PTO} \neq 0$) limit this growth otherwise theoretically to infinity.

3) We seek to determine the solution $\theta(t) = \theta_0\sin(\omega t + \varphi)$. To do this, we introduce this solution into the differential equation obtained in question 2:

$$-\theta_0\omega^2\sin(\omega t + \varphi) + \left\{\frac{C_A + B_{PTO}}{I_y}\right\}\theta_0\omega\cos(\omega t + \varphi) + \omega_R^2\theta_0\sin(\omega t + \varphi) = \frac{\tau_{inc}}{I_y}\sin(\omega t)$$

Grouping the terms gives:

$$\alpha\theta_0\omega\cos(\omega t+\varphi)+\{\omega_R^2-\omega^2\}\theta_0\sin(\omega t+\varphi)=\frac{\tau_{inc}}{I_y}\sin(\omega t)$$

with:

$$\alpha=\frac{C_A+B_{PTO}}{I_y}$$

By expanding the functions $\cos(\omega t+\varphi)$ and $\sin(\omega t+\varphi)$ to only display the functions $\cos(\omega t)$ and $\sin(\omega t)$:

$$\alpha\theta_0\omega\{\cos(\omega t)\cos(\varphi)-\sin(\omega t)\sin(\varphi)\}\\+\{\omega_R^2-\omega^2\}\theta_0\{\sin(\omega t)\cos(\varphi)+\cos(\omega t)\sin(\varphi)\}\\=\frac{\tau_{inc}}{I_y}\sin(\omega t)$$

By grouping the terms proportional to $\cos(\omega t)$ and those proportional to $\sin(\omega t)$, we deduce:

$$\cos(\omega t)\{\sin(\varphi)\{\omega_R^2-\omega^2\}\theta_0+\cos(\varphi)\alpha\theta_0\omega\}\\=\left\{\frac{\tau_{inc}}{I_y}+\alpha\theta_0\omega\sin(\varphi)-\{\omega_R^2-\omega^2\}\theta_0\cos(\varphi)\right\}\sin(\omega t)$$

To verify this equation at any time t, it is necessary that:

$$\frac{\tau_{inc}}{I_y}+\alpha\theta_0\omega\sin(\varphi)-\{\omega_R^2-\omega^2\}\theta_0\cos(\varphi)=0$$

$$\sin(\varphi)\{\omega_R^2-\omega^2\}\theta_0+\cos(\varphi)\alpha\omega\theta_0=0$$

We deduce from the second equation that:

$$tg(\varphi)=\frac{\alpha\omega}{\omega^2-\omega_R^2}$$

The first equation is transformed by successive operations:

$$\frac{\tau_{inc}}{I_y}+\theta_0\cos(\varphi)\{\alpha\omega tg(\varphi)-\{\omega_R^2-\omega^2\}\}=0$$

$$\frac{\tau_{inc}}{I_y} + \theta_0 \cos(\varphi) \left\{ \frac{\alpha^2\omega^2 + \{\omega_R^2 - \omega^2\}^2}{\omega^2 - \omega_R^2} \right\} = 0$$

$$\theta_0 = \frac{\tau_{inc}}{I_y} \frac{1}{\cos(\varphi)} \frac{\omega_R^2 - \omega^2}{\alpha^2\omega^2 + \{\omega_R^2 - \omega^2\}^2}$$

Trigonometry also establishes that:

$$\frac{1}{\cos^2(\varphi)} = 1 + tg^2(\varphi) = 1 + \left(\frac{\alpha\omega}{\omega^2 - \omega_R^2} \right)^2 = \frac{\alpha^2\omega^2 + \{\omega_R^2 - \omega^2\}^2}{\{\omega^2 - \omega_R^2\}^2}$$

but the sign of $\cos(\varphi)$ is to be determined.

Consider the following two cases:

– if $\omega > \omega_R$, then $tg(\varphi) > 0$. By taking $0 < \varphi < \pi/2$, we choose the positive value of $\cos(\varphi)$:

$$\frac{1}{\cos(\varphi)} = \frac{\sqrt{\alpha^2\omega^2 + \{\omega_R^2 - \omega^2\}^2}}{\omega^2 - \omega_R^2}$$

– if $\omega < \omega_R$, then $tg(\varphi) < 0$. By taking $\pi/2 < \varphi < \pi$, we choose the negative value of $\cos(\varphi)$:

$$\frac{1}{\cos(\varphi)} = \frac{\sqrt{\alpha^2\omega^2 + \{\omega_R^2 - \omega^2\}^2}}{\omega^2 - \omega_R^2}$$

Both cases give the same expression of $\cos(\varphi)$. Hence, the general expression for the amplitude of oscillation:

$$\theta_0 = -\frac{\tau_{inc}}{I_y} \frac{1}{\sqrt{\alpha^2\omega^2 + \{\omega_R^2 - \omega^2\}^2}}$$

4) The oscillation of the flap is $\theta(t) = \theta_0 \sin(\omega t + \varphi)$. Let us summarize the results of question 3:

$$tg(\varphi) = \frac{\alpha\omega}{\omega^2 - \omega_R^2}$$

$$\theta_0 = -\frac{\tau_{inc}}{I_y} \frac{1}{\sqrt{\alpha^2\omega^2 + \{\omega_R^2 - \omega^2\}^2}}$$

Then, returning to the expression of the energy recovered during a time interval of duration t established in question 1:

$$E(t) = B_{PTO}\int_0^t \dot{\theta}^2(t')dt' + K_{PTO}\int_0^t \theta(t')\dot{\theta}(t')dt'$$
$$= B_{PTO}\int_0^t \dot{\theta}^2(t')dt' + K_{PTO}\left\{\frac{\theta^2(t)}{2} - \frac{\theta^2(0)}{2}\right\}$$

The average power recovered during a time interval of duration t equal to an integer number of wave periods ($t = nT = n2\pi/\omega$) is written as:

$$\wp_{average} = \frac{E(nT)}{nT} = B_{PTO}\frac{1}{nT}\int_0^{nT} \dot{\theta}^2(t')dt'$$
$$= B_{PTO}\theta_0^2\omega^2\frac{1}{nT}\int_0^{nT} cos^2(\omega t' + \varphi)dt'$$

We then deduce:

$$\wp_{average} = B_{PTO}\theta_0^2\omega^2\frac{1}{nT}\int_0^{nT}\frac{1 + cos2(\omega t' + \varphi)}{2}dt' = B_{PTO}\theta_0^2\omega^2\frac{1}{2}$$

By replacing θ_0 it with its expression, we get:

$$\wp_{average} = B_{PTO}\frac{1}{2}\left(\frac{\tau_{inc}\omega}{I_y}\right)^2\frac{1}{\alpha^2\omega^2 + \{\omega_R^2 - \omega^2\}^2}$$

To optimize recovery, we vary K_{PTO}, which acts on ω_R, and B_{PTO}, which influences α.

The value of ω_R that maximizes the energy recovered is $\omega_R = \omega$. The optimal value of K_{PTO} that adjusts the resonance frequency of the flap to the wave frequency is:

$$K_{PTO,opt} = I_y\omega^2 - g(\rho V - M)d_1$$

In this case:

$$\omega_R = \sqrt{\frac{K_{PTO,opt} + g(\rho V - M)d_1}{I_y}} = \omega$$

The average power recovered is then simplified to:

$$\wp_{average} = B_{PTO}\frac{1}{2}\left(\frac{\tau_{inc}}{I_y}\right)^2\frac{1}{\alpha^2} = \frac{1}{2}\tau_{inc}^2\frac{B_{PTO}}{(C_A + B_{PTO})^2}$$

To maximize the energy recovered by playing on B_{PTO}, it is necessary to determine the value of B_{PTO} such that:

$$\frac{d\wp_{average}}{dB_{PTO}} = 0$$

This is obtained for $B_{PTO,opt} = C_A$.

We deduce the maximum average power recovered by the flap:

$$\wp_{average,max} = \frac{\tau_{inc}^2}{8C_A}$$

5) With the data from the statement:

$$\tau_{inc} = A\rho gHh^2L \qquad C_A = B\rho gh^3L\sqrt{h/g}$$

$$\wp_{average,max} = \frac{\tau_{inc}^2}{8C_A} = \frac{A^2}{8B}\rho g\sqrt{gh}H^2L$$

To evaluate the performance, it is necessary to compare $\wp_{average,max}$ to the available resource in terms of wave energy. It is quantified by the energy flow per unit width (W/m) of a wave crest (see section 8.1.3). It is therefore natural to estimate the efficiency of wave energy recovery by the flap as the ratio of the power recovered to the energy flow of a wave with crest width equal to the panel width L:

$$r_{flap} = \frac{\wp_{average,max}}{\overline{\Im}_{wave}L}$$

For long waves, the group speed is:

$$c_g = c = \sqrt{gh}$$

so that:

$$\overline{\Im}_{wave} = \rho g\frac{H^2}{8}\sqrt{gh}$$

The yield is therefore:

$$r_{flap} = \frac{\wp_{average,max}}{\Im_{wave} L} = \frac{A^2}{B}$$

A wave with period $T = 14$ s propagating in a water layer of depth $h = 8$ m is a long wave. The wavelength and wave number are then given as (section 8.1.2):

$$\lambda = \sqrt{gh}T$$

$$kh = \frac{2\pi h}{\lambda}$$

We verify that $th(kh) \cong kh$, with the data in Table 8.2, to validate the long wave hypothesis.

h	T	λ	kh	$th(kh)$
m	s	m	–	–
8	14	124.02	0.41	0.38

Table 8.2. *Verification that a wave of period T = 14 s propagating in a water layer of depth h = 8 m is a long wave*

6) The forcing applied by the incoming wave is $\tau_{wave}(t) = \tau_{inc} sin(\omega t)$.

The parameter is K_{PTO} chosen so that the resonance angular frequency of the system is equal to the angular frequency of the waves. We also choose B_{PTO} so as to maximize the recovered power. For these optimal values of $B_{PTO,opt}$ and $K_{PTO,opt}$, the phase shift is $\varphi = \pi/2$ (question 3) and:

$$\theta_0 = -\frac{\tau_{inc}}{I_y}\frac{1}{\alpha\omega} = -\frac{\tau_{inc}}{2C_A\omega} = -\frac{AH}{2Bh\omega}\sqrt{\frac{g}{h}}$$

using the data from the statement:

$$\tau_{inc} = A\rho gHh^2L \qquad C_A = B\rho gh^3L\sqrt{h/g}$$

The panel's response is written as $\theta(t) = \theta_0\cos(\omega t)$.

A wave condition is characterized by its crest-to-trough amplitude H and angular frequency ω.

The average power recovered varies with the square of the wave height H^2.

The oscillation angle amplitude of the flap increases proportionally with the wave amplitude H. This amplitude must be controlled. Under certain conditions (see question 7), it will not be possible to operate for resonance conditions.

7) The ratio of the average power recovered to the maximum power recovered is written as:

$$\frac{\wp_{average}(B_{PTO}, K_{PTO})}{\wp_{average,max}} = \frac{B_{PTO}B_{PTO,opt}}{I_y^2}\frac{4\omega^2}{(\omega_R^2 - \omega^2)^2 + \alpha^2\omega^2}$$

We define the variables x and y by:

$$\omega_R^2 = \frac{K_{PTO} + g(\rho V - M)d_1}{I_y} = \omega^2 x$$

$$B_{PTO} = B_{PTO,opt}y$$

The optimized values of the PTO parameters correspond to $x = 1$ and $y = 1$. Using these variables, we write:

$$\omega_R^2 - \omega^2 = \frac{K_{PTO} + g(\rho V - M)d_1}{I_y} - \omega^2 = \omega^2(x - 1)$$

$$\alpha = \frac{B_{PTO,opt} + B_{PTO}}{I_y} = \frac{B_{PTO,opt}}{I_y}(y + 1)$$

The ratio of the average power recovered to the maximum power recovered then becomes:

$$\frac{\wp_{average}(B_{PTO}, K_{PTO})}{\wp_{average,max}} = \frac{4\left(\frac{B_{PTO,opt}}{I_y\omega}\right)^2 y}{(x - 1)^2 + \left(\frac{B_{PTO,opt}}{I_y\omega}\right)^2 (y + 1)^2}$$

The dimensionless parameter:

$$C = \frac{B_{PTO,opt}}{I_y\omega} = B\frac{\rho g h^3 L}{\omega I_y}\sqrt{\frac{h}{g}}$$

governs the variation of the average power recovered with the values of the PTO parameters.

The variation $\theta_O(B_{PTO}, K_{PTO})$ in the angle amplitude of oscillation of the flap is expressed for its part:

$$\theta_0 = -\frac{\tau_{inc}}{I_y} \frac{1}{\sqrt{\alpha^2\omega^2 + \{\omega_R^2 - \omega^2\}^2}} = -A\frac{\rho g H h^2 L}{I_y\omega^2} \frac{1}{\sqrt{C^2(y+1)^2 + (x-1)^2}}$$

The power recovered per unit width of the flap and the angle amplitude of oscillation of the panel are given in Table 8.3 for three configurations which differ by the wave height H and the values of the dimensionless coefficients A and B linking the torques τ_{wave} and τ_{rad} to the wave parameters (equations [8.8] and [8.9] of the statement). The characteristic values of the flap are $I_y = 4 \times 10^6$ kg·m² and $L = 20$ m. The water height is taken at $h = 8$ m and the wave period $T = 14$ s. Figure 8.9 shows the iso-contours of the relative power as a function of the parameters B_{PTO} and K_{PTO}. The iso-contours of the oscillation amplitude θ_0 are superimposed on the graphs.

H	A	B	h	ω	L	θ_0	$\frac{\wp_{average,max}}{L}$	r_{flap}
m	–	–	m	rad/s	m	°	kW/m	–
1	0.2	0.1	8	0.45	20	17.67	4.35	0.4
1	0.4	0.2	8	0.45	20	17.67	8.69	0.8
2	0.4	0.2	8	0.45	20	35.34	34.76	0.8

Table 8.3. *Maximum average power recovered and angle amplitude of flap movements for three wave heights H and different values of the dimensionless coefficients A and B*

The graphs in Figure 8.9 illustrate the phenomenon of resonance. The recovered power is maximum for $K_{PTO} = K_{PTO,opt}$ and $B_{PTO} = B_{PTO,opt}$. The resonance angular frequency of the panel is then equal to the wave angular frequency. We also observe that the variations of the recovered power with the values of the PTO parameters remain quite gentle. Over the considered variation range of PTO parameters, the recovered power is reduced at most to 30% compared to the maximum average power that can be recovered.

The recovered power varies with the square H^2 of the wave height while the angle amplitude θ_0 is proportional to the wave height, as confirmed in Table 8.3. In real maritime conditions, it is necessary to apply active control on the flap. It consists of modifying the PTO parameters according to the height H of incident waves and their angular frequency ω in order to maximize the energy but also to maintain the oscillation of the flap within an acceptable range. For a wave height $H = 1$ m, we can consider choosing the PTO parameters for the resonance conditions. For larger wave heights, the choice of PTO parameters is primarily used to maintain the angle amplitude of oscillation θ_0 at a value below 20°. The limits of use of the linear wave model[8] are another reason to keep the amplitude of oscillation of the flap below 20°.

As the recovered power varies with the square H^2 of wave height, one could think that the sizing of the flap should favor the recovery of the energy of the highest waves. This is not actually the case. The highest waves are rare and the energy recovered over the time they are active is small compared to the energy recovered during a typical year. It is the waves of intermediate heights[9], because their amplitude is quite large and because this wave regime has a higher probability of occurrence, which contribute the most to the recovery of wave energy. For the highest waves, the priority is to limit the angle amplitude of oscillation of the flap, for the reasons previously mentioned and to ensure the survival of the wave energy converter. The forces applied by extreme waves on the wave energy system are considerable.

8 The linear model studied in this exercise does not take into account the vortices which are generated at the edges of the panel at each time the direction of rotation of the flap changes. These vortices apply a nonlinear resistant torque to the panel which varies with the square $\dot{\theta}^2(t)$ of the angular speed of rotation of the flap. This additional resistive torque reduces the recovered power. Maintaining, through the choice of PTO parameters, an oscillation amplitude of less than 20° makes it possible to limit this additional reduction in the recovered power, which is not taken into account in the linear model of this exercise. For more details, please consult Mory et al. (2015).

9 See Mory et al. (2015).

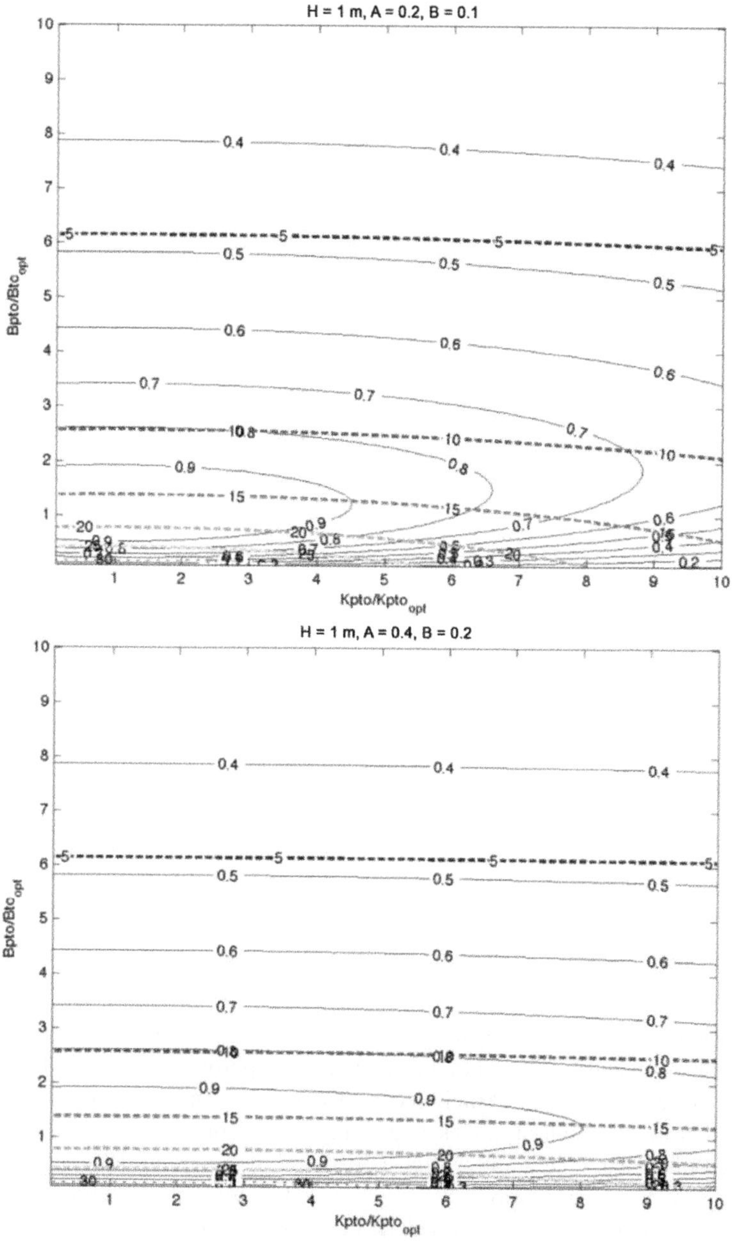

Figure 8.9A. *Iso-contours of the relative power (continuous lines) and iso-contours of the angle amplitude of oscillation (dashed lines)*

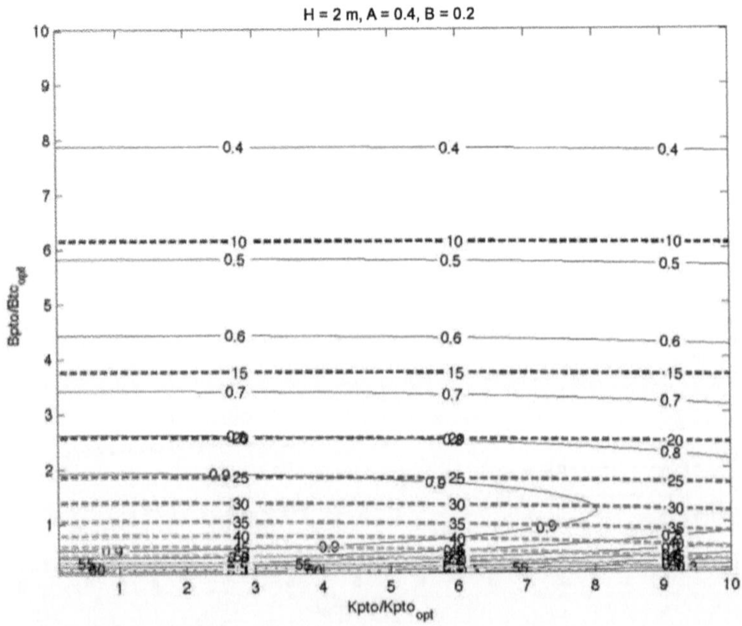

Figure 8.9B. *Iso-contours of the relative power (continuous lines) and iso-contours of the angle amplitude of oscillation (dashed lines) (continued)*

8.3. State of play on wave energy recovery

The wave energy flow increases with the square H^2 of the wave height H, defined as the difference in altitude between crest and trough of the waves. The consequences of this for wave energy recovery are of several orders:

– Sea states are variable and wave energy converters (WEC) must therefore be designed to recover energy under different conditions.

– If we thought that a wave energy converter should primarily recover the energy of the highest waves, this is not in practice the appropriate choice. The highest waves are rare, the duration of their occurrence is short. Therefore, the energy that is recovered from the highest waves is actually small compared to the total energy recovered in a year. The additional costs of building a system designed to recover the energy of the highest waves are not justified in view of the energy recovered. Ultimately, it is the waves of intermediate height which bring the most energy. Their height and the duration of these sea states are the most significant.

There is a wide variety of wave energy converters. We limit our inventory to a few which have given rise to developments at the level of industrial prototypes, distinguishing two categories:

– *Passive WECs*: their operation is not modified over time depending on the height and period of the current wave regime.

– *Active WECs:* their operating parameters are adapted over time to variations in wave height and period.

In section 8.2, we designed a WEC that recovers wave energy using a panel that oscillates under the action of waves. This is an active converter. The exercise highlights several characteristics of this system:

– The PTO device recovers wave energy by adequately opposing the movement of the flap pushed by the waves.

– The system consisting of the oscillating flap and its PTO system has a resonance frequency, which can be modified by changing the PTO parameters. The recovered power is maximized when the resonance frequency is equal to the frequency of waves.

– PTO parameters are modified to adjust the resonant frequency with the frequency of waves. They are also modified to control the amplitude of the oscillation of the panel. Ensuring the survival of the installation in storm conditions is a key issue in the sizing of such an installation.

The exercise in section 8.2 is inspired by the Oyster device shown in Figure 8.10, developed by the Scottish company Aquamarine Power. A first prototype (Oyster 1) had a power of 315 kW. A second industrial prototype (Oyster 800), with a power of 800 kW, was deployed at sea in 2012 and connected to the network. Despite the progress made, the company Aquamarine Power ceased its activity in 2015.

Figure 8.10 shows that the Oyster system has an additional conversion stage. The oscillating movements of the flap, irregular in nature, do not lend themselves to the supply of alternating current at a precise frequency. As a result, the PTO used in the exercise activates a device (double-acting piston), which pressurizes water circulating in a hydraulic loop. Electricity is produced in the classic way, using a hydraulic turbine placed on land at a point in the hydraulic loop. The successive stages of the energy conversion processes of the Oyster system are indicated in Figure 8.10.

Figure 8.10. *Oyster WEC developed by Aquamarine[10], implementing an oscillating panel. For a color version of this figure, see www.iste.co.uk/mory/mechanical.zip*

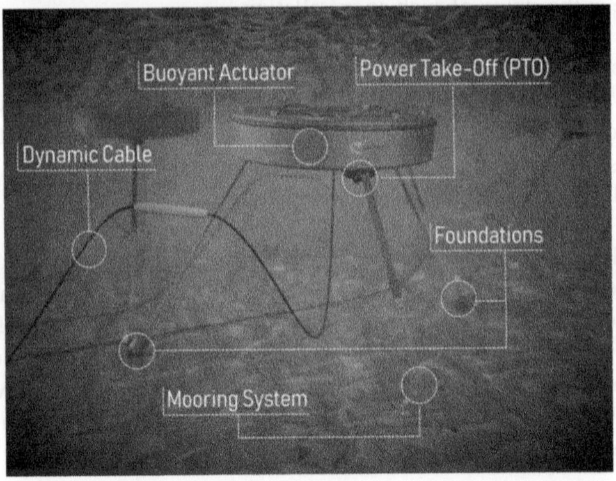

Figure 8.11. *CETO heaving Buoy WEC[11] developed by Carnegie Clean Energy. For a color version of this figure, see www.iste.co.uk/mory/mechanical.zip*

10 Available at: https://www.emec.org.uk/about-us/wave-clients/aquamarine-power/.

11 Available at: https://www.carnegiece.com/ceto-technology/.

Another device, shown in Figure 8.11, consists of using buoys that oscillate vertically under the action of waves. The movement of the buoy activates a PTO. The CETO concept, developed by the Australian company Carnegie Clean Energy, implements this idea. It is similar to the previous device in the sense that it uses the principle of resonance to optimize energy recovery. The two devices differ in that the flap oscillates, pushed by the kinetic energy of the horizontal velocity component produced by the waves while the floating buoy oscillates, pushed by the kinetic energy of the vertical velocity component generated by the wave. The position in the sea of either device must be chosen so that an oscillating flap is placed at a place where the movements produced by the waves are horizontal while the buoy must be placed where the movements produced by the waves are vertical. For example, near a dike which reflects waves (see section 8.1.4), it should be avoided placing a heaving buoy on a node of the standing wave (the vertical velocity component there is zero). On the other hand, an oscillating flap will be well positioned at this point to recover the wave energy (the amplitude of variations of the horizontal velocity component is maximum).

Among the passive converters of wave energy, we show in Figure 8.12 the oscillating water column (OWC) system. It has a cavity leaning against the shore at the base of which the alternating flow and ebb seawater flow enters through a submerged slot. A wave crest raises the water level in the cavity and compresses the air contained in the upper part (dark arrows). The compressed air activates a turbine which produces electricity. The water level falls down with the ebb of the wave, putting the air into depression in the cavity (light gray arrows). The air flow in the turbine then changes direction. The turbines used by OWC systems are Wells turbines (element in dark gray). These turbines have the remarkable property of rotating in the same direction for both directions of air flow.

Figure 8.12 shows the OWC wave energy conversion plant built in Mutriku by the Basque energy agency Ente Vasco de la Energia (EVE). With a total length of 100 m, it has 16 chambers each equipped with a Wells turbine (supplied by Wavegen). The nominal power of the plant is 300 kW. The Mutriku power plant has been connected to the network since 2011 and provides energy of 300 MWh per year. The project was based on a shared cost concept. The sea wall had to be rebuilt to protect the port of Mutriku. The idea was to modify the project to include the wave energy recovery installation. Additional civil engineering costs linked to the electrical plant were €4 million and the equipment for the turbines and electrical installations cost €1.7 million. These costs could have been reduced if the project had included the electricity production plant from the start. This was not the case, generating additional costs and causing difficulties during implementation. The sincerity with which EVE communicated about this project (in articles or during communications at international congresses[12]) must be highlighted. As it stands, a very rough estimate of the production price is €1 per kWh, assuming depreciation over 20 years. It is a high cost, but the project has undeniably made it possible to acquire know-how and evaluate the potential for improvement of the technology.

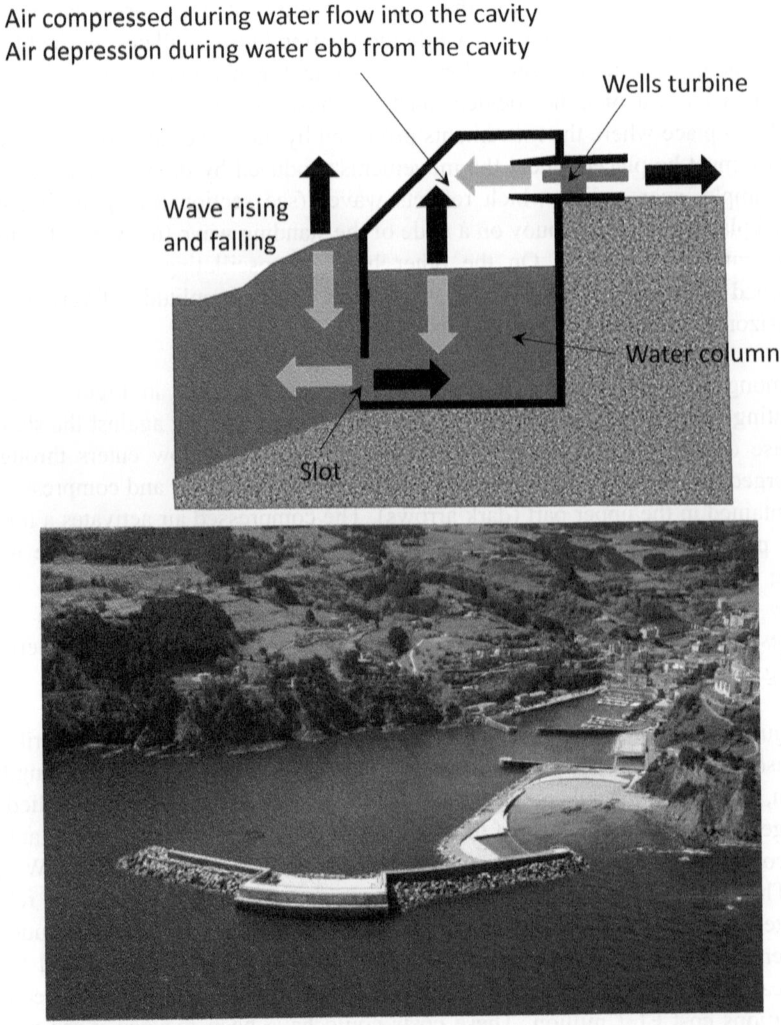

Figure 8.12. *OWC WEC and photograph of the Mutriku Wave Power Plant (Biscay Marine Energy Platform BiMEP). For a color version of this figure, see www.iste.co.uk/mory/mechanical.zip*

12 See Torre-Enciso et al. (2009).

Finally, we show in Figure 8.13 the passive WEC which exploits the overflow of waves on a dike. Its concept consists of equipping a dike with reservoirs distributed at different heights to recover the water brought by the rise of waves on the dike, depending on the height of incident waves. The gravity potential energy of the volume of water recovered in the caissons is then captured by flowing the water back to the sea through low-head turbines. It is a passive WEC. Its optimization consists of limiting the number of reservoirs (there is no point in having a very high reservoir which is empty 99% of the time). The promoters of this technology also defend the problem of shared costs: when a new dike must be built to protect a port, a secondary function of electricity production can be considered, as long as the additional costs for wave energy recovery remain reasonable.

Compared to the considerable energy potential of waves, the powers and energies currently recovered by wave energy systems remain modest. The wave energy sector has been able to develop systems that have reached the stage of industrial prototypes. A few of them were connected to the network. Their economic viability remains fragile.

Figure 8.13. *Sea Slot Cone Generator (SSG) WEC recovering the wave overflow on a sea wall*[13]

13 Source: Vicinanza et al. (2012).

References

ADEME (2011). Dans l'air du temps, l'énergie éolienne. Document No. 6893, Angers.

ADEME (2020). Coûts des énergies renouvelables et de récupération. Données 2019. Report No. 0109895, Angers.

Aelbrecht, D. (2015). What future for tidal power? In *Congrès EWTEC*, Nantes.

Babarit, A. (2017). *Ocean Wave Energy Conversion: Resource, Technologies and Performance.* ISTE Press, London, and Elsevier, Oxford.

BRGM (2003). Dessalement et recharge artificielle : synthèse technico-économique. Final report BRGM/RP-52262-FR.

Haar, L. and Gallagher, J.S. (1978). Thermodynamic properties of ammonia. *Phys. Chem. Ref. Data*, 7(3), 635.

Idel'cik, I.E. (1986). *Mémento des pertes de charge.* Eyrolles, Paris.

Jaumotte, A. and Decock, P. (1992). Aménagements hydroélectriques. *Technique de l'Ingénieur*, B4405.

Landau, L. and Lifchitz, E. (1959). *Fluid Mechanics*. Pergamon Press, Oxford.

Megnint, L., Verdurand, G., Rey, R. (2008). Turbines hydrauliques – Description et fonctionnement. *Technique de l'Ingénieur*, BM 4405.

Mory, M. (2010). *Écoulements pour les procédés.* Hermès-Lavoisier, Paris.

Mory, M., Dugor, J., Baudry, V., Babarit, A., Clément, A.H. (2015). A comparative study of the wave energy collected at Saint-Jean-de-Luz (France) by an overtopping converter and a flap activated converter. In *European Wave and Tidal Conference.* EWTEC, Nantes.

Remigy, J.C. and Desclaux, S. (2007). Filtration membranaire (LI, NF, UF) : présentation des membranes et des modules. *Technique de l'Ingénieur*, J 2791.

Rey, R., Bakir, F., Poulain, J. (2012). Pompes rotodynamiques – Similitude et conception des pompes centrifuges. *Technique de l'Ingénieur*, BM 4303.

SER (2022). Panorama de l'électricité renouvelable au 30 juin 2022. Document, Syndicat des énergies renouvelables.

Torre-Enciso, Y., Ortubia, I., López de Aguileta, L.I., Marqués, J. (2009). Mutriku wave power plant: From the thinking out to the reality. In *European Wave and Tidal Energy Conference*. EWTEC, Uppsala.

Vicinanza, D., Margheritini, L., Kofoed, J.P., Buccino, M. (2012). The SSG wave energy converter: Performance, status and recent developments. *Energies*, 5, 193–226.

Index

A, C

E, F

V, W

Other titles from

iSTE

in

Energy

2023

ALVAREZ-HERAULT Marie-Cécile, GOUIN Victor, CHARDIN-SEGUI Trinidad, MALOT Alain, COIGNARD Jonathan, RAISON Bertrand, COULET Jérôme
Distribution System Planning: Evolution of Methodologies and Digital Tools for Energy Transition

BEN KHADER BOUZID Monia, CHAMPENOIS Gérard
Electrical and Mechanical Fault Diagnosis in Wind Energy Conversion Systems

JANNOT Yves, MOYNE Christian, DEGIOVANNI Alain
Heat Transfer 1: Conduction and Convection
Heat Transfer 2: Radiation and Exchangers

ROBYNS Benoît, LENGLET Claude, BARRY Hervé, BOZZO-REY Malik
Smart Users for Energy and Societal Transition

2022

ABOUELATTA Mohamed, SHAKER Ahmed, GONTRAND Christian
Smart Power Integration

BOUTAUD Benoit
Energy Autonomy: From the Notion to the Concepts

GONTRAND Christian
Electromagnetism: Links to Special Relativity

2021

ROBYNS Benoît, DAVIGNY Arnaud, FRANÇOIS Bruno, HENNETON Antoine, SPROOTEN Jonathan
Electricity Production from Renewable Energies (2nd edition)

2020

BOISGIBAULT Louis, AL KABBANI Fahad
Energy Transition in Metropolises, Rural Areas and Deserts

SOUALHI Abdenour, RAZIK Hubert
Electrical Systems 1: From Diagnosis to Prognosis
Electrical Systems 2: From Diagnosis to Prognosis

2019

BENALLOU Abdelhanine
Energy Transfers by Convection
(Energy Engineering Set – Volume 3)
Energy Transfers by Radiation
(Energy Engineering Set – Volume 4)
Mass Transfers and Physical Data Estimation
(Energy Engineering Set – Volume 5)

LACHAL Bernard
Energy Transition

2018

BENALLOU Abdelhanine
Energy and Mass Transfers: Balance Sheet Approach and Basic Concepts
(Energy Engineering Set – Volume 1)
Energy Transfers by Conduction
(Energy Engineering Set – Volume 2)

JEMEÏ Samir
Hybridization, Diagnostic and Prognostic of Proton Exchange Membrane Fuel Cells: Durability and Reliability

RUFFINE Livio, BROSETA Daniel, DESMEDT Arnaud
Gas Hydrates 2: Geoscience Issues and Potential Industrial Applications

VIALLET Virginie, FLEUTOT Benoit
Inorganic Massive Batteries
(Energy Storage – Batteries and Supercapacitors Set – Volume 4)

2017

BROSETA Daniel, RUFFINE Livio, DESMEDT Arnaud
Gas Hydrates 1: Fundamentals, Characterization and Modeling

LA SCALA Massimo
From Smart Grids to Smart Cities: New Challenges in Optimizing Energy Grids
(Advanced Smart Grids Set – Volume 2)

MOLINA Géraldine, MUSY Marjorie, LEFRANC Margot
Building Professionals Facing the Energy Efficiency Challenge

SIMON Patrice, BROUSSE Thierry, FAVIER Frédéric
Supercapacitors Based on Carbon or Pseudocapacitive Materials
(Energy Storage – Batteries and Supercapacitors Set – Volume 3)

2016

ALLARD Bruno
Power Systems-on-Chip: Practical Aspects of Design

ANDRÉ Michel, SAMARAS Zissis
Energy and Environment
(Research for Innovative Transports Set – Volume 1)

DUFOUR Anthony
Thermochemical Conversion of Biomass for the Production of Energy and Chemicals

SOUSTELLE Michel
Phase Transformations
(Chemical Thermodynamics Set – Volume 5)
Thermodynamics of Surfaces and Capillary Systems
(Chemical Thermodynamics Set – Volume 7)

2015

MONCONDUIT Laure, CROGUENNEC Laurence, DEDRYVÈRE Rémi
Electrodes for Li-ion Batteries: Materials, Mechanisms and Performance
(Energy Storage – Batteries and Supercapacitors Set – Volume 2)

ROBYNS Benoît, FRANÇOIS Bruno, DELILLE Gauthier,
SAUDEMONT Christophe
Energy Storage in Electric Power Grids

ROSSI Carole
Al-based Energetic Nanomaterials: Design, Manufacturing, Properties and Applications
(Nanotechnologies for Energy Recovery Set – Volume 2)

TARASCON Jean-Marie, SIMON Patrice
Electrochemical Energy Storage
(Energy Storage – Batteries and Supercapacitors Set – Volume 1)

SOUSTELLE Michel
Phases Modeling Tools: Application to Gases
(Chemical Thermodynamics Set – Volume 1)
Modeling of Liquid Phases
(Chemical Thermodynamics Set – Volume 2)
Thermodynamic Modeling of Solid Phases
(Chemical Thermodynamics Set – Volume 3)
Chemical Equilibria
(Chemical Thermodynamics Set – Volume 4)
Ionic and Electrochemical Equilibria
(Chemical Thermodynamics Set – Volume 6)

2014

DE LARMINAT Philippe
Climate Change: Identification and Projections

2012

BECKERS Benoit
Solar Energy at Urban Scale

FOULADGER Javad
Electrothermics

2011

GAO Fei, BLUNIER Benjamin, MIRAOUI Abdellatif
Proton Exchange Membrane Fuel Cell Modeling

MONMASSON Eric
Power Electronic Converters: PWM Strategies and Current Control Techniques

MULTON Bernard
Marine Renewable Energy Handbook

RAZIK Hubert
Handbook of Asynchronous Machine with Variable Speed

2010

BRUNET Yves
Energy Storage

JUFER Marcel
Electric Drives: Design Methodology

2009

SABONNADIÈRE Jean-Claude
Low Emission Power Generation Technologies and Energy Management
Renewable Energy Technologies

2008

MEUNIER Gérard
The Finite Element Method for Electromagnetic Modeling

Printed and bound by CPI Group (UK) Ltd, Croydon, CR0 4YY
13/09/2026
14961047-0004